The Tesla F

T0231916

The Tesla Revolution

Why Big Oil is Losing the Energy War

Rembrandt Koppelaar and Willem Middelkoop

AUP

Cover design: Studio Ron van Roon
Lay-out: Crius Group, Hulshout

Amsterdam University Press English-language titles are distributed in the US and Canada by the University of Chicago Press.

ISBN   978 94 6298 206 2
e-ISBN   978 90 4853 195 0 (pdf)
e-ISBN   978 90 4853 196 7 (ePub)
NUR   781 | 961

Table of Contents

Prologue

Nearly 10 years ago, I started investigating the history of oil and how its depletion could impact our postindustrial society. The situation concerned me after I learned about it firsthand from the experienced geologist Colin Campbell (PhD, formerly at BP) who had studied the issue for decades. Would there be enough oil around to fuel society in the centuries or even decades to come?

The book Willem and I coauthored in 2008, *The Permanent Oil Crisis*, emerged from this concern. Our aim was to raise awareness of the issue of peak oil and the need to transition to alternative forms of energy. Its basic premise was that a structural undersupply of cheap oil would disrupt the world economy, because the peak in conventional oil production was at hand. Oil supplies had been flat for several years, and oil prices were on the rise, with limited new sources of supply in sight. The situation was urgent given the lack of alternatives at the time, and we needed to transition to clean sources of energy.

In this new book, *The Tesla Revolution*, we examine what has happened in oil markets since then, with almost 10 years of learning added to the mix. Thanks to large-scale shifts towards alternatives over the last decade, our view is more positive now.

The world of clean energy is advancing rapidly thanks to the efforts of countless people who, like us, are concerned about the end of cheap oil, about how fossil fuel-based carbon emissions are linked to climate change, or both. Just as Nikola Tesla, together with Thomas Edison, revolutionized the world 120 years ago by sparking the electrical age, we are now entering a revolution in clean transport, electricity, and heating.

This new 'Tesla Revolution' in clean energy is driven primarily by the urgency to scale alternatives to oil in transportation, as signaled by high oil prices.

We know that oil is and will continue to be more expensive than in the early 2000s and prior, since we now need to extract either lesser-quality oil, or at more remote locations, or at far greater depths. The recent rise of shale oil in the United States is not changing this situation, since most of it is also costly to extract, as we explore in Chapter 4. Shale oil took the world by storm, and just like the oil industry, we did not see it coming early enough.

The shale oil boom is drying up, however, since the Middle East started pumping flat out in 2014, and oil prices have recently dropped even lower, to $40-$50 per barrel (still three times higher than in the early 2000s and prior). Since March 2015, shale oil production has fallen by 15%–in just 16 months. Since the costs of extraction are high—shale oil is not the cheap oil we're used to—major investments have been scrapped, well drilling has halted, and company bankruptcies are growing.

This situation is not unique to shale oil; if we look at oil production globally, we now need hundreds of billion-dollar investments every year in deep-water fields and oil sands to increase oil production, and many projects have recently been postponed.

Our key message is still valid. We cannot rely on continued smooth growth in (cheap) oil production. There is a large downside risk that oil supply will slump within the next 10 years, bringing substantial economic repercussions as almost all transportation today depends on oil. Not to mention the risk of severe geopolitical instability, since history teaches us that Western countries secure oil supplies via covert operations or military means, which we elaborate on in Chapter 3.

In this post cheap-oil era, we believe we need to work on reducing oil dependency and risk. In a figurative sense, sourcing transportation energy from clean sources where possible is an economic 'pension policy' that anyone who cares about their future should buy into. The faster we can scale alternative energy sources for transportation, the more likely it is that sufficient energy will be available.

The second driver of the Tesla Revolution is the urgency to scale alternatives for all fossil fuels—especially coal—to dramatically reduce carbon emissions. Carbon reductions are needed to halt the chemical alteration of the earth's atmosphere and thereby minimize the disruptive risks of climate change.

Reducing carbon pollution is currently a big driver of many government energy policies and a factor in the investment strategies of many companies and financial institutions. In our view, climate change is just as big a risk as cheap oil depletion in the next decades, and much greater for 2050 and beyond, since increasing climate disruptions will affect the world's economies. We just don't know what the effects will be in our lifetimes, let alone closer to the year 2100 and beyond.

Many people working to transform our fossil fuel-based economy, including the CEOs of Tesla Motors and Toyota, share our concerns. In our view, any book on energy published today needs to look at the extent to which carbon emission reductions are driving the world away from fossil fuels, as we discuss in Chapter 5. We will also update you on last year's Paris Agreement on Climate Change, carbon reduction policies, and Big Oil's underground carbon stock and the investment implications of it.

The key questions we explore are at the heart of coal and natural gas: Has coal begun its long-term demise, as carbon reductions require? Will natural gas play a major role in the future as a bridge to a clean energy world?

That brings me to the impact that the two drivers—expensive oil and climate change—is having on the global energy picture.

In 2008, when writing our previous book, Tesla Motors was a small player with its Roadster electric sports car, and its Chinese counterpart Build Your Dreams (BYD) only sold mini electric city cars. Now, thanks to many innovations, especially in lithium-ion batteries, the entire car world is changing rapidly.

In the wake of Tesla Motors' rise, major car companies around the world, from the United States to Germany to Japan, are aggressively pursuing electric cars, hybrid cars, or fuel-cell car models. Soon, they may follow further in the footsteps of Tesla and BYD which, at the time of publication, are the first two fully integrated electric car—battery—solar-energy companies, bringing renewable driving to your garage or front door.

Thanks to their efforts, it is likely that, not too long from now, car companies will take away transportation market share from Big Oil as electric cars in all segments become both desired and cost competitive. That will not put oil companies out of business (at least not for the foreseeable future), but it will push them to provide oil increasingly for trucking, shipping, flight, and chemicals, which are still more challenging uses to tackle. We examine this and a lot more in Chapters 1 and 6, not just for batteries and electric cars, but also for fuel cells, solar photovoltaic cells, wind power, electricity grids, and other technologies.

So much is happening around the world daily that the news is difficult to keep up with. Did you know that over 40,000 German households have battery systems connected to their solar panels? Or that more than 200 million electric bikes, scooters, and motorcycles are being driven on China's streets already? Or that solar panels power at least 40 million households in regions without electricity grids?

The lack of good information has led to a lot of confusion when it comes to how fast renewables are entering the world's energy system. In our experience, many opinions and so-called facts are vented on the Internet and in newspapers based on outdated information and data, usually limited to experience within the country in which people live, or are ridden with bias. Part of this confusion comes from a lack of up-to-date information on the scale of renewables in the world's energy system as a whole.

To reduce some of this confusion, we paint a picture of the pace of change of the clean energy transition relative to the world's energy system in Chapter 1. Based on the most up-to-date data, up to the end of 2016, we explain how much fossil fuel and clean energy is used, for different energy uses, and the differences across continents and climates.

This provides a bird's-eye view of the current situation seen from the perspective of where we are headed and the scaling that is needed to accelerate the Tesla Revolution—a key part of the book—as we need to know whether the world is moving fast enough and how it could move faster to accelerate the rise of clean energy in the world's energy mix.

We hope that if you aren't already enthusiastic about clean energy, our book will inspire you to join and accelerate the Tesla Revolution. We all have a role to play in bringing the earth closer to a clean energy world. Without people buying electric cars or solar panels, building great tech innovations, shaping energy policies, or financing large-scale clean energy infrastructure, not much will change. We also hope that you find the content useful and that its clarity contributes to your thinking on energy, to help you make better energy decisions in your daily life. We have more decision power than we think.

Rembrandt Koppelaar (rembrandtkoppelaar@gmail.com)
London, November 2016

Special Introduction[1]

This book is about the energy revolution in which Tesla Motors plays a significant role. The name of the company honors Nikola Tesla, one of the greatest engineers and inventors ever.

Who was Nikola Tesla?

Nikola Tesla, born in 1856, was a Serbian-American inventor, electrical engineer, mechanical engineer, physicist, and futurist best known for his contributions to the design of the modern alternating current (AC) electricity supply system.[1]

In 1875, Tesla started at Austrian Polytechnic in Graz, Austria, on a Military Frontier scholarship. Tesla claimed that he worked from 3 a.m. to 11 p.m., including Sundays and holidays. After his father's death in 1879, Tesla found a package of letters from his professors to his father, warning that unless he was removed from the school, Tesla would die from overwork. During his second year, Tesla came into conflict with Professor Poeschl over the Gramme dynamo, when Tesla suggested that commutators were not necessary. At the end of his second year, Tesla lost his scholarship and became addicted to gambling.[2] He never graduated from the university and did not receive grades for the last semester. In December 1878, Tesla left Graz and severed all relations with his family to hide the fact that he had dropped out of school.[3]

---

1   Based on public (Wikipedia) sources.

In 1881, Tesla moved to Budapest to work under Ferenc Puskás at a telegraph company, the Budapest Telephone Exchange. Within a few months Tesla was elevated to the chief electrician position. During his employment, Tesla made many improvements to the central station equipment and claimed to have perfected a telephone repeater or amplifier, which was never patented nor publicly described.[2]

In 1882, Tesla moved to France where he began working for the Continental Edison Company, designing and making improvements to electrical equipment. In June 1884, he emigrated to New York City in the United States.[4] He was hired by Thomas Edison to work at his Edison Machine Works on Manhattan's Lower East Side. Tesla's work for Edison began with simple electrical engineering and quickly progressed to solving more difficult problems.[5]

Tesla was offered the task of completely redesigning the Edison Company's direct current generators. In 1885, he said that he could redesign Edison's inefficient motors and generators, making an improvement in both service and economy. According to Tesla, Edison remarked: 'There's $50,000 in it for you—if you can do it.'[6],[7]

After months of work, Tesla fulfilled the task and inquired about payment. Edison, saying that he had only been joking, replied, 'Tesla, you don't understand our American humor.' Instead, Edison offered a $10 a week raise over Tesla's $18 per week salary. Tesla refused the offer and immediately resigned.[7]

After leaving Edison's company, Tesla partnered with two businessmen in 1886, Robert Lane and Benjamin Vail, who agreed to finance an electric lighting company in Tesla's name, Tesla Electric Light & Manufacturing. The company installed electrical arc light-based illumination systems designed by Tesla. It also designed dynamo electric machine commutators, the first patents issued to Tesla in the United States.[3],[8]

The investors showed little interest in Tesla's ideas for new types of motors and electrical transmission equipment. They were more interested in developing an electrical utility than inventing new systems. They eventually forced Tesla out, leaving him penniless. He even lost control of the patents he had generated, since he had assigned them to the company in lieu of stock. He had to work at various electrical repair jobs and as a ditch digger for $2 a day.[9],[10]

In late 1886, Tesla met Alfred S. Brown, a Western Union superintendent, and New York attorney Charles F. Peck. The two men were experienced in setting up companies and promoting inventions and patents for financial gain. Based on Tesla's patents and other ideas, they agreed to back him financially and handle his patents. Together they formed the Tesla Electric Company in April 1887. They set up a laboratory for Tesla at 89 Liberty Street in Manhattan, where he worked on improving and developing new types of electric motors, generators, and other devices.[9]

In 1887, Tesla developed an induction motor that ran on alternating current, a power system format that was starting to be built in Europe and the United States because of its advantages in long-distance, high-voltage transmission.[11]

In 1888, *Electrical World* magazine editor Thomas Commerford Martin (a friend and publicist) arranged for Tesla to demonstrate his alternating current system, including his induction motor, at the American Institute of Electrical Engineers (now IEEE). Engineers working for the Westinghouse Electric & Manufacturing Company reported to George Westinghouse that Tesla had a viable AC motor and related power system—something for which Westinghouse had been trying to secure patents.[8],[9],[12]

Tesla's demonstration of his induction motor and Westinghouse's subsequent licensing of the patent, both in 1888, put Tesla firmly on the AC side of the War of Currents, an electrical distribution battle being waged between Thomas Edison and George Westinghouse that had been simmering since Westinghouse's first AC system in 1886.[3]

This started out as a competition between rival lighting systems, with Edison holding all the patents for DC and the incandescent light, and Westinghouse using his own patented AC system to power arc lights, as well as incandescent lamps of a slightly different design, to get around the Edison patent.[13]

The acquisition of a feasible AC motor gave Westinghouse a key patent in building a completely integrated AC system, but the financial strain of buying up patents and hiring the engineers needed to build it meant development of Tesla's motor had to be put on hold for a while. The competition resulted in Edison Machine Works pursuing AC development in 1890. By 1892, Thomas Edison was no longer in control of his own company, which was consolidated into the conglomerate General Electric and converting to an AC delivery system at that point.[14]

On 30 July 1891, at the age of 35, Tesla became a naturalized citizen of the United States.[15] He established his South Fifth Avenue laboratory in New York City, and later another at 46 E. Houston Street. He lit electric lamps wirelessly at both locations, demonstrating the potential of wireless power transmission.[2],[3]

Tesla served as a vice president of the American Institute of Electrical Engineers from 1892 to 1894, the forerunner of the modern-day IEEE (along with the Institute of Radio Engineers).[2] Starting in 1894, Tesla began investigating what he referred to as radiant energy of 'invisible' kinds

(X-rays) after he had noticed damaged film in his laboratory in previous experiments.[16]

Much of Tesla's early research—hundreds of invention models, plans, notes, laboratory data, tools, photographs—was lost in the Fifth Avenue laboratory fire of March 1895.[17]

Tesla's theories on the possibility of transmission by radio waves go back as far as lectures and demonstrations in 1893 in St. Louis, Missouri, the Franklin Institute in Philadelphia, Pennsylvania, and the National Electric Light Association.[18]

In 1898, Tesla demonstrated a radio-controlled boat—which he dubbed 'teleautomaton'—to the public during an electrical exhibition at Madison Square Garden. Tesla tried to sell his idea to the US military as a type of radio-controlled torpedo, but they showed little interest.[8],[19]

In 1900, Tesla was granted patents for a 'system of transmitting electrical energy' and 'an electrical transmitter'.

On 6 November 1915, a Reuters news agency report from London erroneously stated that the 1915 Nobel Prize in Physics had been awarded to Thomas Edison and Nikola Tesla. There have been subsequent claims by Tesla biographers that Edison and Tesla were the original recipients and that neither was given the award because of their animosity toward each other; that each sought to minimize the other's achievements and right to win the award; that both refused ever to accept the award if the other received it first; that both rejected any possibility of sharing it; and even that a wealthy Edison refused it to keep Tesla from getting the $20,000 prize money.[7]

In 1928, Tesla received his last patent, US Patent 1,655,114, for a biplane capable of taking off vertically (VTOL aircraft) and then of being 'gradually tilted through manipulation of the elevator devices' in flight until it was flying like a conventional plane. Until his death, he kept working on energy-related inventions including a secret weapon that could send out beams of energy ('death rays'). There are at least 278 patents issued to Tesla in 26 countries that have been accounted for. Many inventions developed by Tesla were not put under patent protection.[7],[20],[21]

Tesla worked every day from 9:00 a.m. until 6:00 p.m. or later, with dinner from exactly 8:10 p.m. in a hotel restaurant. He dined alone, except on the rare occasions when he would give a dinner party to a group to meet his social obligations. Tesla would then resume work, often until 3:00 a.m.[6] He said that he believed that all fundamental laws could be reduced to one. In his article 'A Machine to End War' published in 1937, Tesla stated, 'To me, the universe is simply a great machine, which never came into being and never will end.'[22]

Tesla read and wrote many works, memorized complete books, and supposedly possessed a photographic memory. He was a polyglot, speaking eight languages: Serbo-Croatian, Czech, English, French, German, Hungarian, Italian, and Latin.[6],[7]

Tesla was asocial and prone to isolate himself with his work. However, when he did engage in a social life, many people spoke very positively and admiringly of Tesla. In middle age, Tesla became close friends with Mark Twain; they spent a lot of time together in his lab and elsewhere. Twain notably described Tesla's induction motor invention as 'the most valuable patent since the telephone.'[23],[24]

On 7 January 1943, at the age of 86, Tesla died alone in room 3327 of the New Yorker Hotel, where he had lived for years. Two days later, the FBI ordered the Alien Property Custodian to seize Tesla's belongings, even though Tesla was an American citizen.[17] On 10 January 1943, New York

City mayor Fiorello La Guardia read a eulogy written by Slovene-American author Louis Adamic live over WNYC radio while violin pieces 'Ave Maria' and 'Tamo daleko' were played, while 2,000 people attended the state funeral for Tesla. Despite having sold his AC electricity patents, Tesla was impoverished and in debt when he died.[2],[17]

Tesla Motors' CEO is Elon Musk. Just like Nikola Tesla 100 years earlier, Musk is a remarkable engineer and inventor who will be remembered, just like Tesla, as a man whose ideas have changed the world.

Who is Elon Musk?

Elon Reeve Musk (1971) is a South African-born Canadian-American engineer and inventor. He is the founder and CEO of SpaceX; cofounder and CEO of Tesla Motors; cofounder and chairman of SolarCity; and cofounder of PayPal. Currently (in 2016) he is one of the 100 wealthiest people in the world.

Musk has stated that the goals of SolarCity, Tesla Motors, and SpaceX revolve around his vision to change the world and humanity. His goals include reducing global warming through sustainable energy production and consumption, and reducing the 'risk of human extinction' by 'making life multiplanetary' by setting up a human colony on Mars.[25]–[27] He also has envisioned a high-speed transportation system known as the Hyperloop and has proposed a VTOL supersonic jet aircraft with electric fan propulsion known as the Musk electric jet.[28]

At age 10, he developed an interest in computing with the Commodore VIC-20. He taught himself computer programming and at age 12, sold the code for a BASIC-based video game he created called Blastar to a magazine called *PC and Office Technology* for approximately $500.[29],[30]

Musk was severely bullied throughout his childhood, and he was once hospitalized when a group of boys threw him down a flight of stairs and then beat him until he blacked out. He was initially educated at private schools and moved to Canada in June 1989, just before his 18th birthday, after obtaining Canadian citizenship through his Canadian-born mother.[30]

In 1992, after spending two years at Queen's University, Musk transferred to the University of Pennsylvania, where, at the age of 24, he received a Bachelor of Science degree in physics from its College of Arts and Sciences, and a Bachelor of Science degree in economics from its Wharton School of Business.[31] In 1995, at age 24, Musk moved to California to begin a PhD in applied physics and materials science at Stanford University, but left the program after two days to pursue his entrepreneurial aspirations in the areas of the Internet, renewable energy, and outer space. In 2002, he became a US citizen.[30]

In 1995, Musk and his brother, Kimbal, started Zip2, a web software company, with $28,000 of their father's (Errol Musk) money.[30] The company developed and marketed an Internet 'city guide' for the newspaper publishing industry. Musk obtained contracts with *New York Times* and the *Chicago Tribune*. While at Zip2, Musk wanted to be CEO, but none of the board members would allow it. Compaq acquired Zip2 for $307 million in cash and $34 million in stock options in February 1999. Musk, aged 28, received $22 million from the sale.[32]–[34]

In March 1999, Musk cofounded X.com, an online financial services and e-mail payment company, with $10 million from the sale of Zip2. One year later, the company merged with Confinity, which had a money transfer service called PayPal. The merged company focused on the PayPal service and was renamed PayPal in 2001. PayPal's early growth was driven mainly by a viral marketing campaign where new customers were recruited when they received money through the service.[35] In October 2002, PayPal was acquired by eBay for $1.5 billion in stock, of which Musk received $165 million. Before its sale, Musk, who was the company's largest shareholder, owned 11.7 % of PayPal's shares.[36]–[38]

In 2001, Musk conceptualized 'Mars Oasis', a project to land a miniature experimental greenhouse on Mars containing food crops growing on Martian regolith, in an attempt to rekindle public interest in space exploration.[39],[40] In October 2001, Musk travelled to Moscow to buy refurbished ICBMs that could send the envisioned payloads into space but returned to the United States empty-handed. In February 2002, he was offered a Russian rocket for $8 million. On the flight back from Moscow, Musk realized that he could start a company that could build the affordable rockets he needed.[41] According to early Tesla and SpaceX investor Steve Jurvetson, Musk calculated that the raw materials for a rocket were actually only 3% of the sales price of a rocket at the time. By applying vertical integration and the modular approach from software engineering, SpaceX could cut launch price by a factor of 10 and still enjoy a 70% gross margin. Ultimately, Musk ended up founding SpaceX with the long-term goal of creating a 'true spacefaring civilization.'[42],[43]

With $100 million of his early fortune, Musk founded Space Exploration Technologies, or SpaceX, in June 2002. It develops and manufactures space launch vehicles with a focus on advancing the state of rocket technology.[44] In seven years, SpaceX designed the family of Falcon launch vehicles and the Dragon multipurpose spacecraft. In September 2008, SpaceX's Falcon 1 rocket became the first privately funded kerosene fueled vehicle to put a satellite into Earth's orbit. In May 2012, the SpaceX Dragon vehicle berthed with the ISS, making history as the first commercial company to launch and berth a vehicle to the International Space Station.[45]

In 2006, SpaceX was awarded a contract from NASA to continue the development and testing of the SpaceX Falcon 9 launch vehicle and Dragon spacecraft in order to transport cargo to the International Space Station, followed by a $1.6

billion NASA Commercial Resupply Services program contract on December 23, 2008, for 12 flights of its Falcon 9 rocket and Dragon spacecraft to the Space Station, replacing the US Space Shuttle after it retired in 2011.[46]–[48] Astronaut transport to the ISS is currently handled solely by the Soyuz, but SpaceX is one of two companies awarded a contract by NASA as part of the Commercial Crew Development program, which is intended to develop a US astronaut transport capability by 2018.

In December 2015, SpaceX successfully landed the first stage of its Falcon rocket back at the launch pad. It was the first time in history such a feat had been achieved by an orbital rocket and is a significant step towards rocket reusability, lowering the costs of access to space. This first stage recovery was replicated several times in 2016 by landing on an autonomous spaceport drone ship, an ocean-based recovery platform.[49]–[51]

SpaceX is both the largest private producer of rocket engines in the world and holder of the record for highest thrust-to-weight ratio for any known rocket engine. SpaceX has produced more than 100 operational Merlin 1D engines, currently the world's most powerful for its weight.[52],[53]

His goal is to reduce the cost of human spaceflight by a factor of 10. In a 2011 interview, he said he hopes to send humans to Mars within 10–20 years. In Ashlee Vance's biography of Musk, the entrepreneur reportedly stated that he wants to establish a Mars colony by 2040.[30] SpaceX intends to launch a Dragon spacecraft on a Falcon Heavy in 2018 to soft-land on Mars; this is intended to be the first of a regular cargo mission supply run to Mars, building up to later crewed flights. Musk stated in June 2016 that the first unmanned flight of the larger Mars Colonial

Transporter (MCT) spacecraft is scheduled for departure to the red planet in 2022, to be followed by the first manned MCT Mars flight departing in 2024.[54],[55]

Tesla Motors was incorporated in July 2003 by Martin Eberhard and Marc Tarpenning who financed the company until the Series A round of funding. Both men played active roles in the company's early development prior to Elon Musk's involvement. Musk led the Series A round of investment in February 2004, joining Tesla's board of directors as its chairman. Musk took an active role within the company and oversaw product design at a detailed level but was not deeply involved in day-to-day business operations.[56]

Following the financial crisis in 2008, Musk assumed leadership of the company as CEO and product architect, positions he still holds today, and he owns 22% of the company.[57] In 2014, Musk announced that Tesla Motors will allow its technology patents to be used by anyone in good faith in a bid to entice automobile manufacturers to speed up development of electric cars.[58]

Musk provided the initial concept and financial capital for SolarCity, which was then cofounded in 2006 by his cousins Lyndon and Peter Rive.[59],[60] Musk remains the largest shareholder. SolarCity as of 2016 is the third-largest provider of solar power systems in the United States.[61]

The underlying motivation for funding both SolarCity and Tesla is to help combat global warming. In 2012, Musk announced that SolarCity and Tesla Motors are collaborating to use electric vehicle batteries to smooth the impact of rooftop solar on the power grid. At the moment of writing in 2016, Tesla Motors was in a process to acquire SolarCity,[62],[63] promoting an integrated future, with an electric car producer, a Powerwall-battery maker and a solar roof producing company seamlessly integrated into a new Tesla corporation.

In 2013, Musk unveiled a concept called the Hyperloop: a high-speed transportation system incorporating reduced-pressure tubes in which pressurized capsules ride on an air cushion driven by linear induction motors and air compressors. The mechanism for releasing the concept was an alpha-design document that, in addition to scoping out the technology, outlined a notional route where such a transport system might be built: between the Greater Los Angeles Area and the San Francisco Bay Area.[64]

Chapter 1 – The Tesla Revolution

For the last hundred years, gasoline engines have occupied the mainstream, but if you look forward a hundred years it will not just be gasoline, but diesel, electrics, plug-in hybrids and fuel cell vehicles. We don't yet know which will be chosen.
– Akio Yoyoda, 2016, CEO of Toyota Cars

It's very difficult to make hydrogen, store it, and use it in a car. If you get it from water using electricity it is extremely inefficient; it is about half the efficiency of using electricity directly. It is terrible. The best case hydrogen fuel cell doesn't win against the current case batteries.
– Elon Musk, 2015, CEO of Tesla Motors

All the geniuses here at General Motors kept saying lithium-ion technology is 10 years away, and Toyota agreed with us—and boom, along comes Tesla. So I said, 'How come some tiny little California startup, run by guys who know nothing about the car business, can do this, and we can't?'
– General Motors vice chairman Robert Lutz in 2007, saying that Tesla inspired him to push GM to develop the Chevrolet Volt.

Introduction

In our world, it's all about energy. Cheap electricity has become the true backbone of our economy. We need cheap electricity 24 hours, 7 days a week. Society almost breaks down when power is lost. No TV, Internet, or air conditioning drives most people mad within a few hours. When the grid broke down in New Orleans after hurricane Katrina in 2005, even police were seen looting stores within 24 hours.

The availability of cheap electricity on a continuous basis is taken for granted in high-income countries. Large supply chains of cheap natural gas or coal ensure power plants are able to deliver cheap electricity 24/7. In less-developed countries, people are used to power outages. The majority of factories today operate on a 24/7 basis thanks to the automation of most processes combined with cheap electricity. The more interrupted the flow of electricity is, the more expensive electricity is. The restrictions are so severe that economies without such a constant flow of electricity are not able to expand and grow GDP per capita beyond $10,000–$15,000 per person.[1]–[3]

But now, having burned half of all easily accessible cheap fossil fuels, a strong global movement has emerged. Many politicians, business leaders, and citizens are working towards a 100% renewable energy system. The concerns over climate change are no longer just in the West, but also have become prominent in China and India among other places.[4] In the winter of 2015, one million people marched in the streets during the Paris climate summit calling for a clean energy revolution.

The 178 governments present agreed to move towards low-carbon energy systems to reduce greenhouse gas emissions as fast as possible.[2] Even the Saudi and Russian governments are onboard in the race to low-carbon energy systems. In the words of Vladimir Putin, president of the Russian Federation, 'Our ability to successfully address climate change will determine the quality of life for all people on the planet.'[5],[6] Yes, many of them will even use the climate change threat to be able to pay lip service in international politics.

The transformation of our energy systems is here to stay. It is easy to grasp from the investment numbers in electricity generation systems. In 2015 a total of $329 billion was invested in renewable generation versus only $130 billion allocated to new coal and gas fired power generation.[7],[8] We can also see it in the actions of Big Oil and Big Coal CEOs who have accepted the new reality. While the then CEO of Shell Jeroen van de Veer stated in an interview on television in 2009 that he did not believe in solar energy, saying, 'I need to be older than 100 for solar panels to pay back their investment'[9] today Ben van Beurden, the current CEO, shows how the company has changed its mind: 'I have no hesitation to predict of course that in years to come solar energy will be the dominant backbone of our energy system, certainly of the electricity system.'[10]

In the Middle East, energy tenders have delivered solar energy production streams for $0.03 per kWh, and in Morocco and Texas similar price levels have been reached for onshore wind turbines. Yes, the tipping point has been reached for clean energy. Tesla Motors received sales reservations worth $14 billion worldwide in just a few days for its Model 3 electric car in early 2016.[11] The energy revolution has clearly started.

---

2   The COP21 agreement is anticipated to be ratified and become legally binding in the course of 2017 as over 55 parties with more than 55% of global greenhouse gas emissions will have signed the final agreement by then.

1.   Did Tesla pay you to use this title?

We have no business relation with Tesla, but we decided to use their name for the title of our book because the global transformation is best embodied by Tesla, which itself has gone from being a small electric car company in 2007 to an integrated renewable electricity and mobility business just 10 years later. In our first book on energy, in 2008, we described Tesla as a niche player when it had just launched its first car, the Roadster sports vehicle.

Tesla's corporate mission to accelerate the world's transition to renewable energy is at the heart of this energy revolution. In the words of Musk, 'Since we have to get to a renewable future, it is better to get there as soon as we can.'[12] Tesla can be seen as one of the very few companies today who is at the core of this revolution, just like the bright young engineer Nikola Tesla seemed to be in his time, a good 120 years ago. But let's first point out what the different stages of this revolution are all about:

– Phase 1: The world undergoes the first large growth in the production of clean energy such as wind and solar (2000–2020).
– Phase 2: Growth in global energy needs is now met predominantly by renewable energy sources rather than fossil fuels (2020–2050).
– Phase 3: Clean energy becomes dominant in the energy mix globally, surpassing the amount generated from gas, coal, and oil (2050–2080).
– Phase 4: The end point of the transformation of the global energy system when virtually all energy generated comes from renewable sources complemented with nuclear (2080–2100).

Fig. 1. **Conceptual depiction of the four phases of the Tesla Revolution in energy**

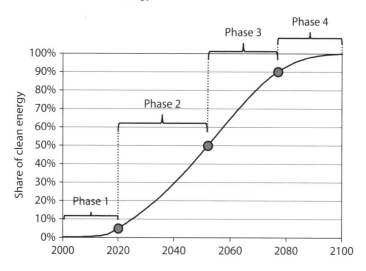

The second phase provides a seminal point in history when growth in energy supply comes primarily from clean renewable sources rather than from fossil fuels. Even though renewables at that point will still be at a relatively small share, it is not to be underestimated as it will fully change the power balance within the energy domain and lead to further acceleration of the already critical mass of people and investments flowing into renewable energy systems.

We expect to reach this phase for all energy uses—transportation, heating, and electricity—within the next decade, and we may have already reached that point for electricity and heating. One could say we are already at a tipping point. Imagine an electric vehicle (EV) being available for the same price as a normal car, but with fewer variable costs for travelling and maintenance. Or a solar-powered electricity system that turns your home into an electricity producer, at a lower cost than incumbent fossil fuels.

2. What makes Tesla such an important part of this revolution?

The founding of Tesla Motors in 2003 and its enormous success are unique in many ways[3] just like the automobile revolution, which started in 1886 when the German engineer Karl Benz invented the internal combustion engine,[4] and was followed by the enormous breakthrough by the American Henry Ford, who launched his car factory in 1903 at the age of 40.[13],[14]

After the discovery of the first 'supergiant' oil field, Spindletop, in Texas in 1901, for the first time millions of liters of oil gushed out every day.[15] Yet unlike in the 2007 movie *There Will be Blood*, nobody knew what to do with such large quantities of oil other than fuelling lamps and the few clunky cars around.[5] It was Henry Ford in the United States who, due to his keen understanding of the market combined with a prowess in technical innovations, reached the mark for the first million cars sold for the first time in history, with his famous Ford Model T. Through the invention of mass production, the car was four times as cheap as its closest competitor, the Oldsmobile, to produce. It could drive over twice as fast (65 km/hour), and handle rural dirt roads with no difficulty. As most Americans lived in rural areas, it was a golden ticket for sales. Car registrations

3   Tesla Motors was incorporated in July 2003 by Martin Eberhard and Marc Tarpenning. Both men played active roles in the company's early development prior to and after Elon Musk's involvement, with Eberhard the original CEO of Tesla until he was asked to resign in August 2007 by the board of directors. Musk led the Series A round of investment in February 2004, investing over seven million personally, before joining Tesla's board of directors as its chairman.

4   Mercedes Benz took a 10% stake in Tesla in 2010 for 50 million USD.[194]

5   A barrel contains 159 litres.

soared from only 2.3% of households owning a car in 1900 to 90% in 1930.[16]

Both Tesla and Ford turned the car industry upside down in just a decade, and both started a revolution way beyond their original idea. The success of Tesla Motors now, just like Ford's in the past, is the result of one great visionary mind who understood that cars could be designed in a totally different way. General Motors vice chairman Robert Lutz has been quoted saying that Tesla inspired him in 2007 to push GM to develop the Chevrolet Volt.

> All the geniuses here at General Motors kept saying lithium-ion technology is 10 years away, and Toyota agreed with us—and boom, along comes Tesla. So I said, 'How come some tiny little California startup, run by guys who know nothing about the car business, can do this, and we can't?'[17],[18]

Just a few years later and all major car companies are rushing to launch their own electric cars as soon as possible. Tesla as a brand and company now stands for the many ingredients that are sparking the revolution in energy worldwide. It focuses on cutting-edge technologies and engineering them to perfection. In doing so, it shows the rest of the world it is possible to make energy production and consumption more sustainable. Elon Musk has stated several times that his goals 'revolve around his vision to change the world and humanity' by rapidly reducing fossil fuel dependence and the impacts of climate change.[19]–[21]

3. What is Tesla's history?

In 2003 Tesla Motors was cofounded by Martin Eberhard and Marc Tarpenning. Both men played active roles in the company's early development prior to and after Elon Musk's involvement, with Eberhard being the original CEO until he was asked to resign in 2007. Eberhard began his career as an electrical engineer and founded NuvoMedia, creators of one of the first e-book readers. He started Tesla because he was passionate about sports cars and concerned about US dependence on imported oil. Elon Musk became involved with Tesla almost from the very start and led the first round of financing in February 2004, investing over $7 million personally, before joining Tesla's board of directors as its chairman. Between 2003 and 2007, total investments in Tesla grew to over $100 million through private financing, in which many famous entrepreneurs participated. The two Google founders were among them.[22]

In early 2008, at the start of the financial crisis, Musk took over as CEO, fired Eberhard and 25% of all Tesla staff, and completed a fifth round of financing ($40 million) to avoid bankruptcy. In that year the company's losses were five times its revenues, and Tesla was in dire shape. By 2009, Tesla had raised a total of almost $200 million and delivered fewer than 150 Roadster sports cars.[23],[24]

Then the tides began to turn. Between 2010 and 2016, it raised another $4.5 billion, of which about half has been spent on new innovations and car designs and the remainder on manufacturing capabilities to ramp up car production and installation of a network of charging stations. This rapid scaling has paid off as Tesla's sales revenue grew from almost nothing to $5 billion.[6]

In June 2010, Tesla Motors launched its initial public offering (IPO), which raised $226 million and made it the first American carmaker to go public since the Ford Motor Company in 1956. Just 13 years after its start, Tesla has grown to a market capitalization of $32 billion, quite close to the $51 billion valuation of its predecessor Ford. The company opened its first Gigafactory for lithium-ion batteries in 2016. Elon Musk has spent an estimated $70 million of his own funds in the development of Tesla Motors and owns almost 30 million Tesla shares, which equates to about 22% of the company in 2016, making him the 37th richest American. Some analysts have shared their doubts about whether Tesla can survive as a stand-alone company because of the huge costs involved in the fast growth of the company.[25],[26]

6   Values based on Bloomberg Financial Data.

4.   What is Elon Musk's goal?

Tesla aims to disrupt the automotive industry by bringing 'many innovative pieces which fit together to bring tremendous advantages' together. Its strategy has been to emulate typical hi-tech life cycles and start with an expensive product for the rich. As the company matures, it is moving 'into larger, more competitive markets at lower price points.' The latest Model 3 car is targeting a mass market with a relatively low starting price of $35,000.[27]

Musk once explained, 'New technology in any field takes a few versions to optimize before reaching the mass market, and in this case it is competing with 150 years and trillions of dollars spent on gasoline cars.'[28] Novel technologies have to be really good to succeed, and building customer relations is key. Tesla is doing this by selling its cars directly to consumers in over 200 company-owned showrooms, 120 of which are outside the US, which is completely different from the standard US dealership model.[7]

It has always been Elon Musk's wish to offer electric cars at prices affordable to the average consumer. Since the launch of the Roadster in 2008, Tesla has sold over 180,000 electric cars worldwide, which is still peanuts compared to almost 69 million cars sold worldwide yearly.[8] The company has promised to build 500,000 cars annually from 2018, after the success of the first sales of Model 3.[29],[30]

Since its founding, the company has shaken up the oil-based car industry by launching four successful 100% electric cars (EVs):

7   A map with all the stores, service centres, superchargers, and destination charging can be found on the tesla website (www.tesla.com) under 'find us'.
8   World car sales grow by 3 million a year at present.

36

- Tesla's first car was a first fully electric sports car, that uses an AC motor descended directly from Nikola Tesla's original 1882 design. The **Roadster** was also the first production car to use lithium-ion battery cells and the first EV with a range of over 200 miles (320 km) per charge. Between 2008 and March 2012, Tesla sold more than 2,250 Roadsters in 31 countries.
- The **Model S** that was launched in 2012 is a luxury five-passenger sedan that holds a range of 390 kilometers (240 miles) per charge. With the same range as the Roaster, it sprints from 0-100 km/h (0-60 miles) in just six seconds. The cheapest Model S was very competitive in the luxury segment, mainly thanks to important tax breaks in the United States, Europe, and Asia. The high-end P85D, with a dual motor, can accelerate to the same speed in just over three seconds (but one has to push the 'insane button' first), making it the fastest sedan on earth, and as fast as a $1 million McLaren super sports car. Over 145,000 Model S cars, which start at $70,000, were sold by the third quarter of 2016. In 2015 it received the US *Consumer Reports* highest score for a car ever, a perfect five-star automobile safety rating, and it is consistently ranks in the top 10 for best environmental performers in the full-size car category.[31]–[34]
- The Tesla **Model X** crossover SUV was launched in 2016 with a 402 kilometers (250 mile) range at a starting price of $100,000, including taxes and excluding credit programs. The roomy car can hold up to seven passengers, is designed with falcon wing doors, can go from 0 to 100 km/h (0–60 mph) within five seconds, and has a towing capacity of 2,268 kilograms. Close to 16,000 were sold by the third quarter of 2016.

– The Tesla **Model 3** four-door compact sedan, to be launched in 2017/2018, is the real breakthrough to mass markets. The Model 3, only 12 cm shorter than the Model S, has the same range and speed and comes with standard autopilot function. It carries a starting price tag of $35,000 (€40,000 if purchased in the euro zone), half the price of a Model S. So it was not surprising that almost 400.000 cars were preordered in just one week in early 2016, representing sales of over $14 billion.[27][9]

9   Pre-order with at least $1000 down payment which are fully refundable when order is cancelled.

5.   Tesla's master plan doesn't stop with cars, does it?

While Tesla started as purely an electric car company, it soon developed into the first fully integrated electric mobility and software company. The software system in all of its cars can be wirelessly updated, like its self-driving autopilot mode. And it has installed a global network of over 800 electric 480 volt Supercharger stations since 2012, so that customers can charge their cars within 20-30 minutes.[35],[36][10] Elon Musk expects future models to reach 500 miles (800 km) on a single charge.[37]

According to Musk, eventually all Supercharger stations will be supplied by solar power. He also promised Tesla owners that use of the network would be free forever,[11] but later explained this would only apply to the luxury S and X models.[38] By 2016, the European Supercharger network reached from Sweden to Spain and Croatia, with extensions to Turkey. Tesla even plans to deploy an India-wide network of Superchargers at the same time as its Model 3 launch.[39],[40] In the United States, Tesla drivers can easily drive, with a few recharges, from coast to coast.

Tesla is also building a $5 billion lithium-ion battery Gigafactory in Nevada, in a business collaboration with Panasonic. Billions of small, cylindrical, lithium-ion cells are to be produced there, similar to those in laptops. The

10   At the end of 2016 the number of Supercharger stations is expected to be around 875. The Supercharger is a proprietary direct current (DC) technology that provides up to 120 kW of power, giving the 90 kWh Model S an additional 170 miles (270 km) of range in about 30 minutes charge and a full charge in around 75 minutes.

11   But Musk's 2012 promise of net-energy-positive solar-powered Supercharger stations has not been met. Only a handful of stations built so far are solar powered.[195],[196] More info at supercharge.info.

factory, by 2018, will churn out 50 GWh in battery capacity per year, sufficient to supply batteries for 500,000 electric cars, thereby doubling global lithium-ion battery production to meet its Model 3 target.[41] And Tesla has already incorporated plans at the same Gigafactory site for a further three-fold expansion. While Tesla's lithium-ion batteries have already dropped from $1,000 per kWh to $200, the Gigafactory is expected to slash costs to below $100 per kWh.[42][12] Battery packs are about a third of Tesla's car costs. At $100 per kWh, the battery in the 55-kWh Model 3 would cost only $5,500.

The company already sells 14 kWh Powerwall battery packs for household storage of electricity and 200 kWh PowerPacks to industrial customers from its Gigafactory.[13] At the time of writing Tesla wanted to merge with SolarCity, operated by Elon's cousin, in order to 'create stunning solar roofs with seamlessly integrated battery storage.' In the words of CEO Elon Musk, 'This made Tesla the world's only vertically integrated energy company offering end-to-end clean energy products.'[43][14] The company announced in June 2014 that it will allow its technology patents to be used by anyone in good faith at zero cost, as part of its aim to accelerate the world of clean technologies.[44]

The Tesla Model S was the first all-electric vehicle fleet to reach the 1 billion electric miles milestone in 2015, closely followed by the Nissan Leaf a month later.[45],[46] In 2014

12   The cost value includes overhead, capital investments and other costs. Many lower cost figures are published that only represent raw material and labour costs.
13   The new 2.0 version of the Powerwall has a battery capacity storage of 14 kWh versus half that for the Powerwall 1.0. Similarly, the new version of the PowerPack holds twice the storage capacity at 200 kWh then the earlier version.
14   The Chinese company Build-Your-Dreams (BYD) actually was the first to combine sales of electric cars, battery solutions, and solar panels within China.

General Motors reported that the plug-in Chevrolet Volt owners had accumulated over 1 billion miles (of which 60% were driven electrically).[47]

At the 2016 'Future Transport Solutions' conference in Oslo, Musk said he expects to produce cars in the future that are even cheaper than the Model 3:[15]

> There will be future cars that will be even more affordable down the road [...] With future generation and smaller cars we'll ultimately be in a position where everyone can afford the car.[48]

6.   Is cheap solar energy the other driver of this revolution?

The biggest game changer in this energy revolution is the fast-declining cost of solar power. Germany now is a central part of this energy revolution, as the second-biggest solar nation in the world. The cost of a 4 kW solar panel system, sufficient for a family household in Western Europe, has dropped from over $22,000 in 2009 to $7,500 by early 2016.[49] This makes a private solar energy system cost-efficient for most households in Germany. Even without subsidies they now pay a lower price for their electricity, a turning point that has been called grid parity. The total cost of a solar panel system is about $0.10 per kWh over its lifetime, versus $0.13 for grid connection and generation costs.[50] When we include high taxes and levies, the comparison becomes even more favorable, as a German household normally pays $0.29 in total for grid-based electricity in 2016.[50] At that price, even batteries to store the energy become attractive, like the popular Sonnen Eco 4 system, which when purchased together with solar panels provides a solar + battery electricity cost of $0.25 per kWh.[51],[52][16]

The cost reductions are even starker for industrial-sized solar parks. Record low costs for utility solar projects in the United Arab Emirates (UAE) have surprised even the strongest sceptics in the world. In November 2014, a 100 MW project was granted to the Saudi Arabian firm ACWA Power

16   The solar and battery cost values are based on total cost including panels, frame, inverter, installation, and maintenance over 25 years, divided by total electricity generated. It does not include loan interest, if a household would fully loan the money at market interest rates of 6%, the cost of a 4 kWh solar panel system would be 14$ cents per kWh, and for a solar + battery system 32$ cents per kWh.

at a price of $0.06 per kWh, far cheaper than the natural gas generation price at $0.09 in the UAE.[53],[54] The record itself was shattered in 2016 when a price of $0.03 per kWh was contracted for the world's largest 800 MW solar plant.[17] This is much cheaper than the first UAE coal power station with a price tag of $0.05 per kWh.[55] Other sunny countries are following close behind such as the United States. Industrial solar projects installed mostly in the sunny Southwest and East of the country costing about $0.11 per kWh in 2013 have dropped to $0.04 in recent tenders.[56],[57] And in India in 2015 a contract was awarded to the Finnish utility company Fortum at a cost of $0.06.[58] So solar parks worldwide are producing energy now at much lower prices than fossil-fuel fed installations.

The large reduction in cost comes from multiple factors. A solar system consists of solar panels, a mounting mechanism, an inverter for DC to AC conversion, cables—and for industrial projects, transmission lines—plus significant 'soft costs' including installation, company overhead, contracting, and so forth. The pure costs of a 1 kW panel alone dropped from $2,055 in 2010 to $687 in 2016.[59] The largest factor has been a massive material cost reduction by reducing the thickness of solar wafers while the amount of light captured has increased substantially. The amount of silicon required per solar wafer has dropped from about 16 grams in 2005 to about 10 grams in 2015.[60][18] The global average efficiency of solar panels grew from around 13% in 2009 to 17% in 2016, with some reaching 20%.[61] The

17   And that record was again shattered with a $2.4 cents per kWh price for a 350 MW solar plant, also in the UAE.[197]
18   Calculation based on a decline in wafer thickness from 285 micrometer in 2005 to 180 micrometer in 2013, a density of silicon of 2.33 grams per cubic centimetre, and a wafer size of 243 cm².

second factor has been a 75% reduction in the price of polysilicon, the raw material that comes from quartz sand used to build solar cells that capture light and transform it into electricity, between 2010 and 2016.[62]

The expectation is that the cost can be reduced by at least another 30% in the next 10 years.[60],[63] Cost could drop even further depending on the success of new forms of solar panel technology. The next big thing is 'tandem cells' that consist of multiple layers, each capturing a different color or emission spectrum of sunlight, thus capturing more energy. Current silicon-based single-layer cell modules cannot capture more than 22% of the energy in sunlight due to physical limits.[64] Yet multilayer tandem cells so far in the lab have achieved close to 50% efficiency, and could in theory go over 80%. SolarCity may soon be one of the first to produce such tandem cells. In its new 1 GW Gigafactory in the United States, in Buffalo, New York, it could start producing 22% + efficiency solar roof tiles in 2017 based on a design where a silicon layer is sandwiched between so-called thin-film layers. SolarCity already builds these at a smaller commercial 32 MW factory in Hangzhou, China.[65]–[68]

7. Will electric cars really take over before 2030?

The world is awash with cars, with over 1.2 billion on the world's roads.[19] In 2015 no fewer than 69 million were sold versus 45 million just 10 years earlier, as car sales ramped up in China from 4 to 21 million in 10 years' time.[30] In contrast, sales in the world's rich OECD countries remained stable at around 30 million per year, with even the United States steady at 7.7 million except for a temporary drop due to the 2008-2009 financial crisis.

Almost two decades have passed since Toyota launched the world's first mass-produced hybrid vehicle, the Prius, in 1997. By 2015, the hybrid vehicle market had grown to 2 million units. When looking at electric vehicles, sales still look pale in comparison with gasoline and diesel powered cars. In 2015 the total number sold was 550,000, including all-electric and plug-in hybrids, which brought the on-the-road total to over 1 million for the first time in history.[69] According to a report by Goldman Sachs in 2016, the EV market could grow to over 2 million cars in 2020 and 4 million in 2025. Hybrid sales could rise to almost 14 million units. By 2030 those numbers could reach 10 million (EV) and perhaps 20 million (hybrid), by the combined forces of car companies and a boost from government incentives.

While innovation in battery technologies is proceeding at a rapid pace worldwide, battery development is booming in Asia in particular. Asian battery makers currently have around 50 GWh of production capacity, equivalent to 50% or more of global output (consumer electronics/automotive

---

19 The total number of passenger cars on the road was estimated at 950 million, plus an additional 250 million light duty vehicles.

battery capacity combined), and have monopolistic shares of 50%-80% of core component materials such as cathode materials and separators. Asian producers have also taken the lead in the development of next generation batteries. Japan accounted for 53% of patent applications filed in 2002–2011, followed by the US at 13%, Europe at 12%, Korea at 10% and China at 8%.

Germany's transport minister Alexander Dobrindt recently rolled out a plan to bring 6 million EVs, or 10% of all German cars, onto the road by 2030.[70],[71] The plan was kick-started by a $4,500 subsidy plus motor vehicle tax exemption for up to 400,000 new cars sold in the next few years, combined with a rollout of 15,000 charging stations.[72] The country is actually one of the latest, as China, the United States, France, Japan, and many others already provide EV tax credits or exemptions, typically in the range of $5,000 to $10,000, and have subsidized charging points.[69] China alone has provided subsidies so far of $5.6 billion, with another $10 billion expected in the next few years.[73] This massive financial stimulus is part of an ambitious target of no fewer than 5 million electric vehicles on the road by 2020.[20] Politicians in Norway and India are even bolder as they are working on policies to get 100% of cars sold by 2030 to be all-electric.[74],[75]

So now we can see multigovernment efforts to rapidly scale up electric vehicles. They are united via the 15-country global Electric Vehicles Initiative (EVI), which covers 90% of all car markets. Jointly, their ambitious aim is to sell 6 million EVs per year by 2020 with 20 million on the road.[76] For that target to happen, all current plans for scale-up from

20   In 2015 EV sales in China were estimated at 200,000 when not including fake sales.[198]

all electric car providers have to be achieved (table 1).[77] Even reaching half that target will lead to a new phase of this revolution, where global annual growth in additional passenger vehicles sold of 3 million is fully met by electric cars.

Likely that tipping point will be reached by 2022, as electric cars without any subsidies or tax credits will then cost the same as equivalent internal combustion cars, according to Bloomberg New Energy Finance.[78] Electric cars then win out financially on cost savings. The key to this will be the launch of models that boast both range and low cost, as mass-market all-electric cars need at least a 200-mile range and a price far below $50,000. Today the soon to be sold GM Chevrolet Bolt E3 and Tesla Motors Model 3 have a 200+ mile capability at a $30,000 to $35,000 starting sales price. Nissan and Renault has announced that they will upgrade their Leaf and Zoe models for similar performance in 2017. Also Ford, and Hyundai made announcements for this mileage range and cost level, yet none have released any model details.[79],[80] Several new entrants are joining this electric car race. Apple's secret car project might also be just such a car, which according to rumors will be launched in 2020.[81] The UK electronics company Dyson as well as US car company Fisker also made up-beat announcements about their electric cars with highly advanced batteries that are coming.[82],[83]

Long-term forecasts have been made by Bloomberg New Energy Finance, which project that 20% of all vehicles sold by 2030 will be EVs and 35% by 2040, which translates to 41 million sold annually.[84] If true, about a quarter of all cars on the road will then be electric vehicles worldwide. The International Energy Agency has worked out a similar scenario; they think the total electric vehicle stock could

Fig. 2. Evolution of the global electric car stock: history and projections by Bloomberg and the IEA

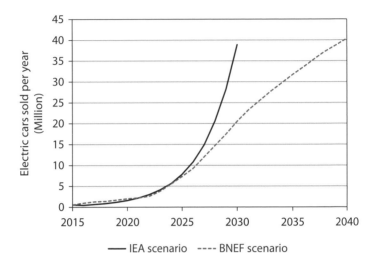

reach 140 million by 2030 and 900 million by 2050, which would mean that two-fifths of the 2 billion passenger vehicles by then will be electric.[69] It is important to note that the conservative IEA is known to historically have been too pessimistic, at least in their projections for solar panels and wind energy sales.

Table 1.   Overview of electric car targets across the world

Company	2015 Car sales[85]	Of which all-electric & plug-in[86]	Current models mid-2016	Announced target
Renault-Nissan*	2.8 million (Renault) 6,1 million (Nissan)	106,000	Nissan Leaf, e-NV200, Renault Zoe, Mitsubishi Outlander PHEV	Anticipates 10% of its sales to be plug-in and all-electric in 2020.[87],[88]
Build Your Dreams (BYD)	62,000	62,000	Qin, Tan, E6,Denza EV	To double EV production every year up to 2018 to 0.5 million cars sold by 2018.[89],[90]
Tesla Motors	52,000	52,000	Model S, X, and 3	An 'impossible' target of 0.5 million sold per year by 2018.[91]
Volkswagen Group (includes Audi)	10 million	49,300	e-Golf, Audi R8 e-tron, VW e-UP, Golf GTE, Passat GTE	Volkswagen aims to push sales to 2-3 million electric cars by 2025.[92] Its subsidiary Audi wants to make 25% of US sales to be plug-in or all-electric vehicles by 2025.[93]
Zhejiang Geely** & Khandi	532,000 (Geely) 28,000 (Khandi)	28,400	Geely Emgrand Khandi Panda EV, Khandi K10 EV	To switch 90% of its 450,000 annual Chinese car sales to electric by 2020.[94]
BMW	1.9 million	28,000	I3, I8, X5 40e	Announced a 100,000 sales per year goal for 2020.[95]

Company	2015 Car sales[85]	Of which all-electric & plug-in[86]	Current models mid-2016	Announced target
Ford	6.6 million	21,100	Focus Electric, Fusion Energi, C-mAx Energi	Seeks to invest $4.5 billion in electric vehicles by 2020 with 40% of its sales options electric.[96]
General Motors	9.8 million	17,500	Chevrolet Volt, Chevrolet Bolt E3, Cadillac CT6 hybrid	Aims for 500,000 plug-in electric vehicles sold by 2017.[97]
BAIC	524,000	17,400	Beijing E-series	Anticipates its EV sales may reach 700,000 units by 2020.[98]
Volvo	503,000	10,200	S60 Plug-In, XC90 T8	To push 10% of global car sales into the electric segment by 2020.[99]
Hyundai	7.9 million	7,500	Kia Soul EV	Aims to expand its electric car range by several vehicles by 2020.[100]
Honda	4.7 million	110	Fit EV, Accord Plug-in	Announced that two-thirds of sales should be hybrids, plug-in, and all-electric by 2030.[101]

* Includes figures for Mitsubishi as Nissan holds a majority stake in the company as of 2016

** Chinese company, which owns Volvo sales since 2010. Figures exclude Volvo sales. The company has a strategic partnership with Khandi as of 2016 for the production of electric vehicles.

8.   Why is the price of electric cars declining so rapidly?

The first electric cars were expensive. The electric vehicle (the EV1) General Motors built in 1996 was a two-seater all-electric car based on lead-acid batteries, with a $53,000 price tag (inflation adjusted) and a range of only 120 kilometers. With only 5,000 leased over time, GM crushed them at end-of-lease in a PR blunder of massive proportions.[102],[103]

Tesla Motors' cars are based on lithium-ion batteries, a brilliant strategic decision. Lead-acid batteries were far cheaper, yet costs had matured with no improvements in sight. But lithium-ion batteries were just starting to get better and cheaper. The Roadster came with a price tag of $109,000, of which battery costs alone were over $50,000. In 2016 the same battery pack costs $10,000.[104]–[106][21]

Thanks to smarter material composition and design, the battery pack of the Tesla Model 3 is about half the weight of the Roadster's batteries, but it carries the same energy load.[22] So the 'power per unit weight and time' is increasing at an enormous speed. In 2016 Tesla can produce a battery

---

21   The average lithium-ion battery pack today costs 390 USD per kWh, whilst Tesla has reduced its costs to 190 USD/kWh, and General Motors & LG-Chem costs are estimated at 215 USD/kWh.[106] These values are for the entire battery pack, and not the cells alone.

22   Estimates for the Roadster storage capacity across its warranty life-cycle is 117 Wh per kg, the Model S at 150 Wh per kg, and the Model 3 will likely come close to 200 Wh per kg. These are for the entire battery-packs, not the cells only which come in much higher.

pack with a range to 350 kilometers for $35,000 (figure 3),[23] but expects these costs to halve in a few years.[24]

The rapid adoption of electric vehicles and hybrids over the next decade will spur an enormous increase in the market for batteries. In 2016 Goldman Sachs forecast a six-fold increase in the market for EV batteries, to $24 billion in 2025, from 58 GWh to 387 GWh (demand for automotive batteries rising from 15 GWh to 279 GWh, from a $4 billion to $24 billion market).

The increase in energy density in batteries has been enormous. In 2005 the energy density was 100 Wh/kg; in 2015 over 200 Wh/kg; 300 Wh/kg is expected by 2020; and over 500 Wh/kg in 2025. Experts say this will likely require a transition to a 'post-lithium-ion' technology because we are close to the practical maximum for this type of chemistry.[107] Asian battery makers, responsible for over 50% of global battery output, have taken the lead in the development of next generation batteries, called all-solid-state batteries, which include lithium-oxygen and lithium-sulfur, which can hold at least five times more energy.[108],[109][25,26]

Goldman Sachs analysts think these will enter the market in the 2025–2030 time frame, resulting in a further reduction

23   Other drivers of cost declines include smarter and more streamlined manufacturing, economies of scale in purchasing contracts, and lower costs of risks and financing. Lithium prices as one of the few commodities were quite stable in recent years and thus did not have a significant impact on cost declines (or increases).
24   The Tesla 3 is sold for $35,000 and is said to have a range of about 300 km.
25   The lithium-ion battery today has a liquid electrolyte between the cathode and anode. An all-solid-state battery has only a solid electrolyte between the cathode and anode and contains no liquid.
26   Japan accounted for 53% of patent applications filed in 2002-2011, followed by the US at 13%, Europe at 12%, Korea at 10% and China at 8% (Goldman Sachs repor, October 2016).

Fig. 3. The cost of batteries in dollars versus gravimetric density (in weight) from 2006 to 2015 and Goldman Sachs forecasts to 2025.[7],[69],[104],[105],[111],[112]

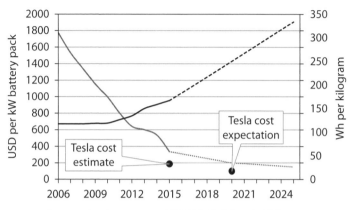

— Avg. market battery cost USD/Wh (left hand side)
⋯⋯ Battery Cost Extrapolation (Goldman Sachs)
— Battery density Wh/kg for cars on market (right hand side)
--- Battery density extrapolation (Goldman Sachs)

by then to $5,000 for a Model 3 type battery pack.[105][27] That would make the cost of an EV fully equivalent to a gasoline car, and cheaper in places where electricity prices are low. Tesla thinks this will already be the case by 2020.[110],[111]

27 Based on an energy density by weight of 500 Wh per kilogram and a battery cost of 100 USD per kWh.

9. How do developments in fuel cell cars compare?

Since the General Motors Electro van was built in 1966 as the first fuel cell car prototype, a lot of automakers have invested in this technology. Fuel cell cars operate by a chemical process that converts fuel into electricity using a cell based on positive and negative charges. The most common is the solid oxide fuel cell (SOFC), which can run on hydrogen that together with oxygen is turned into electricity and water.

Despite dozens of companies building several cars with this design between 1990 and 2010, none made it to the commercial stage until recently. Now four companies have fuel cell vehicles on the market:

- In 2013 **Hyundai** began production of the **hybrid Tucson Fuel Cell Crossover SUV** vehicle which can travel 426 km (265 miles) and can be bought for a $77,000 price tag. Sales have been disappointing with only a few hundred sold by the start of 2016.[113]
- Not long afterwards—in 2015—**Toyota** launched the **subcompact Mirai** that has a range of 502 km (312 miles) and is sold for $57,400 in Japan and the United States and the EU, in areas with hydrogen fuelling infrastructure. So far only a handful of vehicles have been sold.[114]
- **Honda's fuel cell vehicle Clarity**, a six-seater sedan with a range of 700 km (435 miles), is to be sold from 2017 for $63,000 in Japan.[115]
- Daimler is starting sales of its **Mercedes-Benz GLC F-Cell Crossover SUV** in 2017 at $76,000. It is a hybrid combined lithium-ion battery–fuel cell vehicle with a battery-only range of 50 km (31 miles) and a fuel cell range of 500 km (310 miles).[116],[117]

Other carmakers are more cautious, with GM aiming for 2020 and Nissan for 2021 to market their first fuel cell hydrogen car.[118]–[120] The most ambitious so far is Toyota which aims to sell 30,000 fuel cell vehicles per year by 2020, which is challenging given the lack of hydrogen charging infrastructure with only about 80 hydrogen fuelling stations in Japan and 48 in California at the time of writing.[121],[122]

Large-scale production is necessary to bring down the cost by 20%–30% to make these cars affordable. Another challenge lies in the cost of producing and distributing hydrogen. During these steps about 35%-40% of the energy value is lost plus another 50% when converting hydrogen into electricity in the car to drive it.[123],[124] As a consequence, the biggest reason to turn to hydrogen—its much higher energy content when compressed than gasoline or diesel—fades away.

10. What makes the Tesla Revolution so different from previous energy transformations?

The world in the last 200 years has gone through several complete transformations in its energy systems, from wood to coal, from coal to hydro to oil, and finally to gas and nuclear. Pessimists argue that it will take more than 100 years for renewables to replace the incumbent fossil fuel sources. The optimistic perspective goes that the world can be transformed into a 100% renewables society within a matter of 15 to 30 years, as expressed in 2008 for the United States by former vice president Al Gore:

> Today I challenge our nation to commit to producing 100 percent of our electricity from renewable energy and truly clean carbon-free sources within 10 years. This goal is achievable, affordable and transformative.[125]

The pessimistic argument goes that historically, energy transitions were all upgrades in energy content of stored fuels[28], while solar and wind are intermittent sources requiring storage that makes them costly. And we use so much more energy than in the past, when fossil fuels took between 35 and 55 years to go from a 5% to a 25% share in the global energy mix.[126],[127] Optimists often base their thinking on a technology push, looking at computer microprocessor speed improvements doubling every 18 months in the last few decades as predicted by Moore's law. Unfortunately, this cannot be translated directly to the energy domain, where technologies move far more slowly.[128]

---

28   Natural Gas and Oil contain about 3 times as much energy as wood, and coal contains about twice the energy content of wood.

While interesting as frames of thinking, such comparisons are incomplete as several factors make the Tesla Revolution unique relative to past energy transitions: technology transfers, decentralization, government commitment, and innovation speed.

Technology transfer - Today's world is interconnected and technology transfer between countries happens fast. It only took China 10 years to catch up technologically with EU and US wind turbine manufacturers. How? China's largest turbine producer, Goldwind, started in 1997 by licensing 600 kW turbines from Jacobs Energie of Germany. The next step it took was to manufacture advanced 1.2 MW gearless turbines from a design by the Swiss company Vensys. History was made in 2008 when Goldwind acquired a 70% stake in Vensys, giving Goldwind access to intellectual property and Vensys to financial capital.

Today Goldwind is the largest wind turbine manufacturer in the world, and it is also on the forefront of research as it rolls out its 6 MW gearless offshore turbine.[129]–[132] Since mid-2014 Tesla is helping to create new success stories for transportation like Goldwind in electricity, as Tesla has made all their electric car patents available for free with no royalties and no licensing fees.[133] Toyota and Ford are now also allowing competitors to use their electric and fuel cell vehicle patents, but only under a licensing contract for limited duration.[134]–[136]

Decentralization - The decentralized, modular nature of wind and solar energy generation makes installation and expansion fast. While the average time to build a natural gas, coal, or nuclear power plant is around two, four, and seven years respectively, it takes only one year to build a wind and solar park.[137] No additional fuel supply infrastructure and contractual arrangements are necessary, as no uranium, gas, or coal need to be extracted, upgraded, and imported, also cutting down time needed. Communities and even individuals can also partake in the investment unlike previous transitions. In leading solar nation Germany, from 15% to 40% of solar panels sold every year are purchased by households, with the remainder by bigger institutions and utility companies.[138]

Government commitments – A large number of governments today are committed to quickly growing renewable energy and electric cars, driven by energy security and climate change concerns. This commitment means generous financial support in the tens of billions, as well as regulatory support to prioritize clean energy sources that drives hundreds of billions of investments. No such support was present in past transitions. Also the increasing streamlining and specialization between nations for particular technologies is leading to more rapid advances, as opposed to individual countries going at it on their own.

Innovation speed – Today thousands of networks of investors, government R&D organizations, companies, and entrepreneurs exist in large economies. These spheres within countries were much more separate before the 1950s and often relied on only a handful of people. There was a large risk that efforts would come to a standstill for years or decades for lack of funding, ageing, or other reasons. The first commercial industrial-scale solar heat plants for pumping water at 30,000-acre plantations, unbeknownst to many, were built in the 1910s by a German engineer called Frank Schuman, in Egypt and Sudan. At the outbreak of the First World War, the project team abandoned the project, and Schuman died at the end of the war, which stopped all progress, until efforts were restarted 60 years later in Italy and the United States.[139]

11.   Have we already entered an era moving away from fossil fuels?

With little mention in the world news, the last three years were record breaking for electricity generation. Growth in electricity has been dominated by wind, hydro, and solar energy sources in all of the last three years (table 2). We actually have already entered the second phase in the Tesla Revolution for electricity in which growth is dominated by clean energy. Heroic policy and investment in dozens of countries have made this possible—such as Germany, China, the United States, Denmark, Portugal, Kenya, and Uruguay. And remarkably, in 2015, fossil fuel use for electricity saw a monumental decline caused largely by reductions in US coal usage, while solar and wind overtook all other sources.

The picture is still not so favorable for oil use and mobility development. Despite big efforts by Tesla and others, clean energy growth has not yet picked up in the world's transport picture. Globally, total liquid fuel production including all types of oil and biofuels increased by 2.6% and 3.2% in 2014 and 2015, respectively, to a daily output of 91.7 million barrels. Biofuels contributed less than 1/100th to that growth with production rising to 1.5 million barrels per day in 2015. Also, if we look at passenger vehicles, out of the 69 million cars sold in 2015, electric cars and plug-in hybrids contributed less than 1/100th to this total with about 550,000 such cars sold. No other significant alternatives have yet been introduced into the transport market, such as hydrogen, methanol, and ammonia.

The third energy pillar - heating - shows a strong renewable share, yet not so much improvement recently. We are already using 36% renewables for global heat generation,

Table 2. Global electricity generation by source from from 2010 to 2015 in Tera-Watt hours (TWh).[140]-[142]

Values in TWh	2010	2011	2012	2013	2014	2015	Difference 2014-2015
Total electricity	21,494	22,185	22,753	23,336	23,764	23,963	199
Fossil Fuel	14,509	15,116	15,535	15,779	15,913	15,827	- 86
Clean Energy	6,985	7,069	7,218	7,557	7,850	8,136	285
Per source:							
Coal	8,659	9,138	9,158	9,568	9,726	9,401	- 325
Natural Gas	4,813	4,887	5,100	5,066	5,154	5,313	159
Fuel oil	970	1,066	1,138	1,145	1,034	1,113	79
Nuclear	2,768	2,653	2,472	2,493	2,543	2,577	34
Hydro	3,466	3,516	3,693	3,822	3,908	3,946	38
Solar	33	64	102	143	191	253	62
Wind	342	436	526	644	716	841	125
Other renewables	376	400	425	456	492	518	26

Table 3. Global heat generation by source from 2010 to 2015 in gigajoules (GJ) (includes conversion losses in combusting fuels & geothermal)[142]–[149]

Values in GJ	2010	2011	2012	2013	2014	2015	Difference 2014-2015
Total Heat	128,635	130,979	133,567	137,225	137,674	139,050	1,377
Fossil Fuels	83,831	85,372	86,616	87,694	88,925	89,128	203
Clean Energy	44,803	45,607	46,952	49,532	48,748	49,922	1,174
Per source:							
Coal*	27,083	28,348	29,338	30,096	29,916	28,658	- 1258
Natural Gas	40,860	41,296	41,231	41,845	42,539	43,481	942
Oil Products	15,888	15,728	16,047	15,752	16,470	16,989	519
Biomass**	44,003	44,667	45,865	48,303	47,412	48,506	1095
Geothermal	62	77	92	95	103	106	3
Nuclear	27	28	28	28	28	28	0
Solar Thermal	7111	835	9677	1,1066	1,206	1,283	77

* Coal includes peat

** Biomass includes modern and traditional wood fuels, residues, biogas, and municipal waste

mainly from biomass, and we are well into the second phase of the Tesla Revolution. Yet fossil fuel use is still growing, albeit recently more slowly than renewables. The slow-down in fossil fuel growth is caused by declining coal use for heating in the last three years, in contrast to continued growth in natural gas combustion, as well as oil product use in kerosene lamps and fuel oil boilers. If we look at renewables, solar thermal is growing steadily but is still far too small to make a dent. And the share of biomass, composed primarily of traditional burning in small stoves and fires, has remained mostly stable.

12. How much energy does the world need?

The development of the world economy goes hand in hand with increases in energy consumption as transport, electricity, and heating make the modern world work. Global economic growth of 32% from 2004 to 2014 led to an increase of 25% in world energy consumption.[149],[150] Future needs for energy thus depend on how big (or small) the economy will become combined with the efficiency at which we can produce goods and services.

The typical starting assumption is to take a high 3% to 5% economic growth rate and extrapolate this several decades into the future. Energy agencies such as the IEA and the US EIA as well as oil and gas companies like ExxonMobil and BP do this to calculate energy needs. The economic forecast is translated into energy demand by assuming future energy intensities (energy per unit of economic output). In the World Energy Outlook 2016, the IEA assumes a growth rate of 3.5% to 2040, which implies that the world economy will be 2.5 bigger in 25 years' time. And it estimates that this will lead to 48% higher energy consumption by 2050 as energy intensity drops by nearly half. In other words, one unit of energy leads to double economic output by 2040.[151]–[153]

The biggest uncertainty is the economic growth scenario. After the start of the credit crisis in 2007/08, demand for energy first declined in the advanced economies, then slowly started to grow again in 2010, although at a slower pace. The effect was massive as global energy demand dropped 1.6% from 2008 to 2009, and is now growing by a full 1% less than before the credit crisis hit (as economic growth is slower). Today, total energy use in the United States, Germany, France, and Italy is still not as high as it was in 2008.[149]

Fig. 4. Economic growth forecasts, energy use per dollar output, and energy consumption to 2040 from BP, ExxonMobil, and the IEA.[151]–[153]

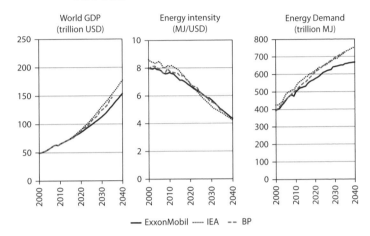

If another debt bubble bursts and economies decline, all current energy scenarios can be put in the garbage bin.

We can also look at this from a country perspective. We could say there are three different types of economies:

– The '**low- and lower-middle income**' countries,[29] with a population of about 3.6 billion people, that use about 12% of all energy used. Their infrastructure is limited, with fewer than 50 cars per 1,000 people, and annual electricity generation is below 1,500 kWh per person, including power needs for production. Often these countries are still largely agriculturally rooted like Nepal, Ghana, and Bangladesh.

29 The definitions for income ranking are made by the World Bank, where low-income is defined for the 2017 fiscal year as economies where people have a gross national income of 1025 USD or less USD per person, lower middle income range between 1026 to 4035 USD, higher middle income between 4036 and 12475 USD, and high income above 12476 USD per person.

- A group of '**upper-middle income**' countries, with a population of about 2.4 billion people, that use about 35% of all energy used. These emerging economies, like Mexico, Thailand, and China, have a rapidly expanding infrastructure and a dominant manufacturing sector in the economy. Car ownership ranges between 50 and 400 cars per 1,000 people, and annual electricity generation lies between 1,500 to 5,000 kWh per person.
- The set of '**higher-income**' countries, with a population of about 1.4 billion people, that use about 53% of all energy used. Their infrastructure is fully established on all fronts, and they rely heavily on services in addition to manufacturing, agriculture, and extraction, like the United States, Germany, and South Korea. Car ownership ranges between 400 and 800 per 1,000 people, and electricity generation from 5,000 to 20,000 kWh per person.

Each of these types has substantially different starting points in terms of infrastructure and energy usage. A general rule of thumb is that three to four times more energy is used and infrastructure built when a country jumps to a higher income group. To give everyone at least an upper-middle-income lifestyle, energy use would need to grow by 30%, and to achieve a higher income lifestyle for everyone, a tripling in total energy use is needed.

The recent example of China which super-charged from a lower- to an upper-middle income country demonstrates the impact. China is now the largest energy consuming country in the world, as energy use tripled between 2000 and 2015. In that period it built 1.23 million km of paved roads, 58,600 km of railways, nearly 1,000 GW of power plant capacity (more than what is installed in the entire European

Union's 28 countries), and millions of new buildings from 2005 to 2015.[154]–[157] Since 2013, however, China's energy use growth has begun to slow down because steel and cement output stagnated at 70 and 200 million tons per month.[158],[159] As the country's infrastructure is nearly built out, it no longer needs as many material inputs, and energy use growth will be much slower than in the past, but the country needs much more energy than before to maintain and use its infrastructure.

China's story could also happen elsewhere—in India, Africa, and Latin America—albeit perhaps at a slower pace. Today India's population of 1.26 billion citizens owns 24 million cars, but oil (or electricity) demand will grow massively if it grows to 600 million cars within the next 30 to 50 years. And many other countries are working on big infrastructure acceleration that together would easily form a 'third China', like Indonesia (258 million people), Bangladesh (161 million), Nigeria (182 million), Ethiopia (99 million).

13.　From fossil fuels to alternatives, what are the scenarios?

We use fossil energy for a lot of different things, from vehicles to electrical appliances, to heating and cooking, to melting iron in large furnaces. Although lots of clean energy transition scenarios are possible, the main ones are listed below. Their impact was calculated from world energy use in 2015 based on the world energy overview we created in the appendix on page 257.

The easy 'big impact' switches to grow renewables use include:

– **Electric cars and motorcycles**, as motorcycles using electricity are cost and mileage effective, and electric cars increasingly so. If a full switch is made globally, oil use would be reduced by 31%, while electricity use would grow by 24%.
– **Wind- and solar-based electricity backed up by gas**, as the cost of wind and solar panels will continue to drop, and for the time being natural gas can fill the gaps during the day when wind and sun are not producing. About 77% of coal and 7% of oil use can be eliminated when replaced by solar and wind electricity, if they grow ten-fold from today's levels.
– **Home and commercial sector heating to solar**, as 85% of the world's population lives in sunny countries. Solar can replace large portions of natural gas, oil, and coal used today in heating boilers and lamps. If two-thirds could be switched to solar thermal, nearly 10% of natural gas use, 4% of oil use, and 3% of coal use would be eliminated.
– **Home cooking to efficient biomass and induction**, as most biomass use is still inefficient and modern biomass

technology could cut use up to 28% (or free up this bio-mass for more people). Also, richer countries can shift to electricity via induction cooking, as induction is nearly 90% efficient. If all homes cooking with natural gas or coal would switch to electricity, coal and gas use would drop by 5% and 1%, against 7% electricity generation growth (or double today's wind and solar).

Several shifts are difficult to make because alternatives are still costly and technically challenging. Some of the difficult to anticipate scenarios are:

– **Shipping**, as limited alternatives exist for the 5% of oil used in long-distance goods transport, with only renewables-based hydrogen as a potential (still too costly today) alternative. The 1% of oil used in inland waters can increasingly use electricity.
– **Trucking**, as the heavy load diesel engine efficiency is challenging to replace. The best alternative today is eHighways with overhead wires for short-distance electric trucking. About 12% of oil is used for the 80 million heavy trucks on the road today.
– **Airplanes**, as planes use minimal weight to fly high in the sky over vast distances. Batteries are too heavy, and so are hydrogen storage tanks today to make hydrogen powered fuel cells work. The 7% of oil used for flying will not go away anytime soon.
– **Industrial energy use**, since for such uses really high temperatures of several hundred to over 1,000 degrees Celsius are needed, and electricity for industrial use typically costs $0.05 or less.

14.   So Big Oil is losing the energy war?

In the aftermath of the oil price spike of 2008, a remark-able development is changing the face of the oil industry. Nearly all the oil and gas majors have started to have severe difficulties replenishing their oil reserves, and production for several is declining. The eight international majors[30] from North America and Europe, which have traditionally dominated oil markets, saw their production drop by 10% to 20%, except for ExxonMobil and Chevron, and reserves for the future were barely maintained. Their share in world oil production is now a mere 12%, down from 17% just 10 years ago. At the same time, exploration and development costs have increased for these majors from below $10 per barrel to at least $30 per barrel. Investment in the oil ma-jors is increasingly less attractive, with return on capital invested dropping from 10%+ to about 5%. In 2015, Total, Shell, and Statoil even had stunning negative capital returns to stockholders.[31]

In a bid to prolong fossil fuel growth in the face of their limited long-term prospects for oil, the eight corporations have embarked on a gas-driven strategy. In the UK, Statoil launched a multimillion dollar campaign called 'Fueling the Future' to boost the image of natural gas. ExxonMobil sent the message that the company produced more electricity from natural gas than ever before as part of its 'Energy Lives Here' campaign. And Total is 'committed to better energy' as a global campaign, of which natural gas is a key part. In

---

30   The group includes BP, Chevron, ConocoPhillips, ENI, ExxonMobil, Shell, Statoil, and TOTAL.
31   Values based on Bloomberg Finance company balance sheet figures.

the last three years, ConocoPhillips, Eni, Shell, Total, and ExxonMobil also produced equal or more gas than oil.

The strategy is unlikely to be sufficient, except if natural gas becomes the dominant source of energy. Car companies and governments are now investing heavily in alternative vehicles (See Q7). Once phase 2 in our energy revolution is reached in the 2020s, in which growth in car use is met predominantly by electric cars, growth in oil consumption will slow down, and we may even reach 'peak oil demand.' And while lots of electricity will be needed in the future, it is more likely that natural gas will be used increasingly to complement solar and wind and to balance out daily fluctuations, as opposed to being the major power source.

That the eight oil majors are not planning for much other than natural gas becomes quite clear from their investments in research. Between 2010 and 2015, a record $500 billion of the high oil price profits were spent on share buy-backs and dividend payouts to investors. Total research spending, in comparison, was only $33 billion, or about 1.5% of gross profits, of which the majority was spent on oil and gas. Just a doubling of spending on clean energy research would make a vast difference. The rapid reduction in solar electricity costs in the last five years came on the back of total global corporate and government research spending of $29 billion. [24]

15. What needs to be done to accelerate the Tesla Revolution?

Imagine a world in the year 2050 where clean energy dominates the energy mix. How do current trends need to accelerate to reach this third phase in the Tesla Revolution? What scaling is needed in project investment, jobs, and research spending in clean energy in the shift from fossil fuels to clean energy sources?

In order for electric vehicles to form the majority of all cars on the road by 2050, sales need to grow from today's 0.6 million to about 15 million by 2030, close to 50 million by 2040, and 100 million by 2050, also assuming growth to 2 billion cars. At this pace, about two out of three cars sold by 2050 will be electric. Since the manufacturing capacity is already there in the car industry, existing factories would need to be retooled from combustion to battery based cars.

The biggest challenges are the large expansion in battery production and the need to provide input minerals lithium and cobalt. At this rate of expansion, three times more lithium needs to be mined by 2030 than today, and 25 times more by 2050.[162] Cobalt production would need to double by 2030, and grow six-fold by 2050.[163]–[165] Large investments in lithium mining as well as battery recycling are needed, as otherwise we will run out of known lithium in the ground at the end of this century.,[166],[167] Current known cobalt reserves would run out within a few decades at this pace of growth, with limited respite from the wider resource base.[168][32] Substituting cobalt in batteries is thus

32   Today about 124,000 tonnes of cobalt are produced, versus 7.1 million tonnes of reserves. About 6 kg of cobalt per electric car is needed for batteries, which implies 600,000 tonnes to produce 100 million electric cars. Some room for reserve expansion exists, with 25 million tonnes identified resources, but it is not a lot.

essential for the long-term success of electric cars, unless a virtuous battery recycling chain with low losses can be established.[169],[170]

If we want to get to a world with a 50%+ share of clean energy in the electricity mix globally before 2030, and 75% by 2050, we need 10 and 20 times the installed number of solar panels, and 5 and 10 times the installed wind power capacity existing today. Even, if we assume the majority of coal and oil generation is phased out, electricity production would still double in this conservative scenario, while natural gas stays at its current share in electricity generation.[33] The jump needed to make this happen is actually quite low. Current global manufacturing capacity for solar panels would need to quadruple in the next 15 years, and windmill production triple, and both would have to stay at those levels until 2050.

In a conservative cost assumption—a 20% and 30% cost reduction for wind and solar by 2050 respectively, plus cost reductions due to greater solar panel output—annual investments for the scenario above need to approximately expand 2.5 times by 2030 in wind and solar and remain at that level. In 2015 we spent $270 billion on installing solar and wind power, or 0.4% of the size of the global economy in gross domestic product (GDP), and $125 billion on fossil fuel generation.[171] In comparison, spending on fossil fuel

33 Assumptions include a 30% and 35% increase by 2050 in nuclear power and hydro-power. Concentrated solar power, geothermal, and bio-energy were deliberately kept at the same level as today. For gas to stay at its current share total capacity needs to expand from 1500 to 2600 GW, resulting in 2 units of gas for every 5 units of solar and wind for balancing, because of which limited battery capacity for electricity storage would be needed. The replacement of all existing and future power plants was taken into account based on a lifespan of 25 years for solar and wind technologies, and 50, 40, 40, and 30 years for coal, natural gas, fuel oil, and nuclear generation respectively.

Fig. 5. Electricity generation scenario with 75% clean energy mix share by 2050

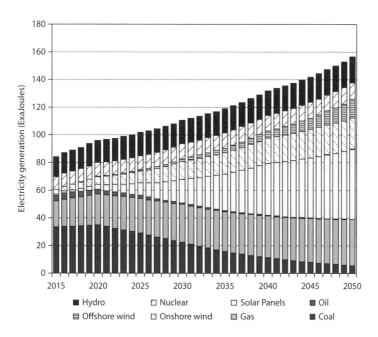

generation in the 1970s was around 0.6% of global economic wealth, and since the 1980s close to 0.2%.[34] The additional costs we need to bear to accelerate the transition are thus high but not extraordinary. Fossil fuel savings are not even factored in, on which we spent at minimum $600 billion in 2015 for electricity generation.[35]

The biggest materials impact of this electricity switch would be on silver, since its use for solar panels needs to

34 Based on historic capital investment values traced back to the 1970s, annual capacity realised for the different fossil fuels, and adjustments for inflation using 2010 real USD value.

35 An average global oil price of 40 USD per barrel, a coal price of 35 USD per ton, and a natural gas price of 5 USD per million BTU.

quadruple by 2030 if no further efficiencies are made. Solar panels today comprise 8% of global silver demand, so a lot more silver mining will be needed. Input needs for other materials like steel, aluminium, glass, concrete, and copper are relatively small compared to global production, and silicon is easily scalable. Jobs in solar and wind manufacturing and installation would need to increase from about 4.4 million today to 12 million by 2030, and 20 million by 2050, which also is within the realm of possibilities.

On top of the investments in expanding capacity, we need to work harder on cracking the tough nuts. How can we replace natural gas use? Make battery electricity solutions for the grid for 24-hour periods more affordable? Improve solar panel efficiency from 17% today to over 30% in the next decades? Create a technically workable solution for seasonal and day-to-day storage? Find fuel alternatives for trucking, shipping, and flight? Provide industrial heat from clean energy? Replace not so abundant materials like silver? Make biofuels from non-food crops available at commercial scale?

The amount of research spending and time to lead to successive breakthroughs for each of these questions will be tens of billions of dollars over decades. It took $45 billion in research spending to achieve the rapid improvements in solar panels from 2005 to 2015, and $16 billion for wind energy. Just for wind and solar we thus needed about $6 billion per year to get to where we are. Today total government energy research spending is close to $19 billion per year,[36] with about three times as much corporate spending.[172]

36 The level is similar to that of the 1980s during the oil crises. The big difference with the early 1980s is that over half of it then was spent on nuclear energy, which today is the case for clean energy.

We have 20 to 30 energy domains that we need to sink money into, resulting in a total spending need of $160 to $240 billion per year to make rapid progress in most of these fields.[172]–[174]

Enhanced space for medium-sized energy tech companies to innovate and explore new options is essential. In this effort we need to have a willingness to 'waste' money on technology firms at levels of technology demonstration to commerciality, since for every one company that is a success about 10 others will go bankrupt. Larger collaboration in research via targeted $100+ million funding sums and programs—instead of multibillion dollar 'moon-shot programs' like nuclear fusion, and thousands of tiny few-million-dollar research grants—would also help.

16.   Are governments serious enough about this Revolution?

In late 2015 the most remarkable announcement at the International Climate Negotiations in Paris did not come from a politician. It was billionaire philanthropist Bill Gates who announced that he has brought 21 governments together to commit to doubling their clean energy research investment by 2021, from $15 to $30 billion per year, working together under a program called Mission Innovation (see mission-innovation.net).[37] He also announced a 'breakthrough energy coalition' of 20+ billionaires that together will invest billions per year in a new clean energy fund to accelerate new energy technologies.[175]

Pushing governments to commit is key as ministers typically operate in response to markets. After the oil crises of the 1970s and 1980s waned, government energy research spending dropped to rock-bottom levels in nearly all countries except the United States and Japan.[176] The abysmal drop in the UK was from $1.7 billion to just $0.1 billion which was only partly reversed as oil prices rose in the last decade.[38] That drop led to the end of a nuclear industry in the UK.

Government policies are also based on 'what can be done' within a short three- to six-year election time frame to score political points. The idea of the 'hydrogen economy' was in the air in the early 2000s in the European Union

37   The most remarkable commitments came from China and the United States, which commit to doubling their research spending to 7.6 and 12.8 billion, respectively, according to mission innovation figures.

38   The ending of government research spending began with the second Margaret Thatcher government in 1987, and the recent increase was set under the first Tony Blair government.

countries. A hydrogen fuel cell program was announced that aimed at 0.5 to 1 million hydrogen fuel cell cars by 2020, based on spending of just up to $100 million per year without any realistic checks on cost, the massive research spending budget required to back it up, or sufficient support of car companies.[177] New upbeat roadmap targets of 2.5 million vehicles were even set in 2008.[178] Not surprisingly, years later none of it has happened, as technologies can't be pushed to affordable cost in just a couple of years, let alone without any significant investments.

A serious energy policy at its heart has a framework with long-term ambitious targets, intermediary financial support before commercialization, sizeable budgets for research to drive down costs, and market regulation that allows for clean energy takeoff. The best regulations so far include a guaranteed price level paid out of electricity bills (also called feed-in tariffs) and capacity tenders where companies bid to build a specified amount of renewables, with a 20-year guarantee attached. Both work because they reduce investment risk and thus can, even at low returns, be effective.

Few governments have managed to bring multidecade continuity to energy policy; notable exceptions include Germany, China, Brazil, and the United States. Though often praised for its world leading wind industry, Denmark is not one of them, as its success came to a standstill in the 2000s. Conservative-liberal parties were elected, privatized the energy sector, and reduced clean energy investment to a minimum. Nothing happened with wind power for 10 years except for wind industry exports, until elections favored the social democrats in 2011 who reinstated wind support and rolled out a plan to cover all energy needs with clean energy in 2050.[179] Yet their influence did not last for long, and today the conservative-liberals are back in power, axing support for clean energy.[180]

The key problem that governments need to become serious about solving is the sudden disruption of a supportive low-risk environment because that disruption kills investment and leads to bankruptcies. Any changes need to be made with incremental adjustments, as Germany has done in reducing the feed-in tariff levels in steps, as opposed to just killing them off in one go. Key to success in parliamentary democracies is the commitment even of opposition parties to long-term energy policies, market support, and research investments. Otherwise at each election cycle this revolution comes to a standstill.

17.   Is this energy Revolution possible without subsidies?

Yes, it's true, the rise of renewables and the success of Tesla were driven to a great extent by government subsidies. Financial support in over 120 countries totals over $135 billion, of which 37% for solar, 27% for wind, 13% for biomass, and 22% for biofuels projects.[153]

The majority of these subsidies are provided in a few countries (Germany, the United States, Italy, Spain, China, Japan, the UK, France, and India). In 2016 the Chinese government gave a subsidy for solar electricity generation of $0.12 per kWh generated, versus an industrial and residential electricity price of $0.15 and $0.08.[181],[182] Since 2015, renewable energy subsidies are being scaled back in several of these countries, with recent large reductions in Spain, Italy, UK, Germany, and the United States, which substantially reduces further expansion of alternative energy projects in these countries.[183],[184]

Yes, the level of renewable subsidies is still substantial, but in comparison they are still much lower than global subsidies for oil, gas, and fossil fuel electricity. Government subsidies to lower the cost of oil products amounted in 2014 to $267 billion, $107 billion for natural gas, and $117 billion for fossil-fuel powered electricity.[185][39] The vast majority of these are provided in Saudi Arabia, Iran, India, Russia, Venezuela, Egypt, Indonesia, China, Algeria, and the United Arab Emirates.

---

39   In a number of reports on also tax credits are included as subsidies, which automatically results in much larger fossil subsidy figures. However, these in essence are reduced revenue intakes.

However, this is far from the complete picture since oil and gas extraction businesses have to pay a part of their profits to the governments of the countries they operate in. This varies between lows of 30% to 40%, in countries like Ireland, Peru, and Morocco, to highs of 80% to 90% in Libya, China, Nigeria, Venezuela, and Iran.[186] The total government income from oil alone in 2014 was estimated at $1.85 trillion in 2014[40], of which the biggest receivers were Saudi Arabia ($290 billion), Russia ($258 billion), the United States ($132 billion), Iran ($100 billion), China ($94 billion), Iraq ($92 billion), Kuwait ($87 billion), and the United Arab Emirates ($76 billion).[187]

This government take comes from royalties, corporate taxation, and shareholder dividend payment tax. In the United States, a company has to pay 12.5% to 30% in royalties based on its contract and a 35% corporate income tax (varying per state), as well as dividend taxes.[186],[188] A big unanswered question of the energy transition is how the government portion of oil and gas extraction income can be replaced in the long term by other sources of income.

Elon Musk's companies (Tesla Motors, SolarCity, and SpaceX) together have benefited from almost $5 billion in government support from a variety of government incentives including grants, tax breaks, factory construction, and discounted loans.[189][41] The companies employ over 20,000 people and operate factories and facilities in California, Michigan, New York, Nevada, and Texas. According to research by the *Los Angeles Times*;

40   Income net of oil extraction costs.
41   Musk owns 27% of Tesla and 23% of SolarCity

The $1.3 billion in benefits for Tesla's Nevada battery factory resulted from a year of hardball negotiations. They shored up the deal with an agreement to give Tesla $195 million in transferable tax credits, which the automaker could sell for upfront cash [...]Tesla buyers also get a $7,500 federal income tax credit and a $2,500 rebate from the state of California. Tesla buyers have qualified for an estimated $284 million in federal tax incentives and collected more than $38 million in California rebates. California legislators recently passed a law, which has not yet taken effect, calling for income limits on electric car buyers seeking the state's $2,500 subsidy. Tesla owners have an average household income of about $320,000, according to Strategic Visions, an auto industry research firm.[189]

In a response, Elon Musk stated:

'If I cared about subsidies, I would have entered the oil and gas industry',[190] and pointed out that Tesla competes with a mature auto industry that has seen massive federal bailouts for General Motors and Chrysler.

18.   Why are low-income countries potentially winners?

The big advantage of the Tesla Revolution is its decentralized, 'modular' nature. Worldwide, individuals and communities increasingly have access to solar panels and windmills at affordable prices with many benefits. Especially solar power is proving to be a game changer, as it is available year round for 85% of the world's population in nearly all of Asia, Africa, Central, South America, and North America excluding Canada.[42]

Governments and companies in many low- and lower-middle-income countries struggle to ramp up centralized fossil fuel energy infrastructure in an urbanizing world. Several hundred million to a few billion US dollars are required to set up one or more fossil fuel power plants with connected grid infrastructure. Substantial planning expertise is needed, and the ability to attract loans from financial institutions can be substantial hurdles for low-income countries. Because of this, about 1.2 out of 3.6 billion people in low- and lower-middle-income countries still have no access to electricity, and globally about 2.8 billion people suffer from costly power outages on a weekly basis in countries such as Tajikistan, Ghana, and Bangladesh.[1] Costly large diesel generator backups are normally set up as stopgaps to deal with the situation. Car fuel gas stations are limited to key roads in rural settings. And the majority of fuels, if not all, are typically imported, making energy supply a very costly affair.

---

42 An imaginary horizontal line drawn on the globe around a latitude of 40 divides countries with plenty of sunshine all year round to those who lack sun in winters. Countries below the line have a sunshine level of 1700 kWh per m² per year or greater.

Solar off the grid is rapidly filling some of the gap by providing electricity to nearly 50 million people in 2015 to those who lived without it, paid out of their own pockets with small loans.[191] The biggest success comes from a combination of 'pico-solar' solutions up to 10 watt, super-efficient appliances, and pay-as-you-go mobile money finance.[43] One of the largest pico-solar providers, M-Kopa in Kenya, has so far leased 400,000 of their 8-watt systems to households at a cost of $0.50 per day for a year, with 1 million units in expected sales by 2017.[191],[192] The solar system is far more affordable than kerosene lamps for lighting or diesel generators for electricity. It also comes with a small battery, which after sunset enables eight hours of lighting and mobile phone charging. It can also power superefficient radios, fans, and a small television. The market is becoming so lucrative that energy giants like Engie (formerly GDF-Suez) invested $20 million in BBOXX, a Kenyan off-grid solar provider, in August 2016.[193] Now for the first time hundreds of millions of households can have light after the sun goes down and extend learning, work, and play hours, and billions can cut costs and get more reliable power.

43   Mobile phone money transfer and wallet systems give rural households access to finance as they do not have access typically to bank accounts. The 20 million users M-Pesa system launched in 2007 by Safaricom is one of the biggest.[199]

Chapter 2 – A History of Fossil Fuel Dominance

No matter what we attempt to do, no matter to what fields we turn our efforts, we are dependent on power. We have to evolve means of obtaining energy from stores which are forever inexhaustible, to perfect methods which do not imply consumption and waste of any material whatever... If we use fuel to get our power, we are living on our capital and exhausting it rapidly. This method is barbarous and wasteful and will have to be stopped in the interest of coming generations.
– Nikola Tesla 1897 and 1915

The transition to widespread use of solar energy has already begun. Our task is to speed it along. True energy security—in both price and supply—can come only from the development of solar and renewable technologies.
– Jimmy Carter 1979, former US president

If you don't invest in oil and gas, you will see more than $200 per barrel.
– OPEC secretary-general Abdalla El-Badri, 2015

Introduction

The Industrial Revolution began in the 1870s with the inventions of the steam engine, electricity, and oil drilling. At that time, virtually everyone alive lived without electricity, traveled primarily by horse, and heated their homes with wood. Life was hard and most time and income were spent on producing and buying food. The inventions of electricity generation and lighting by Nikola Tesla and Thomas Edison changed everything.[1] The era of electricity started with the opening, in 1897, of the first multimegawatt Niagara Falls (waterfall) hydroelectric power plant. Tesla's alternating current (AC) generators enabled the harnessing of the power of falling water and distribution of the electricity to the city of New York. The current AC motor used by Tesla in all of its cars is built using the same laws of physics. We now live in an era of energy abundance. Never before in the history of humankind has so much energy been available. It is difficult to fathom that—just in the last 300 years—we have used more energy than in all the millions of years of human history prior. Let's have a look at what has happened to the world of energy in the last 150 years.

19. So the first Tesla Revolution started a good 100 years ago?

Within a few decades after the first hydroelectric power plant was built at Niagara Falls, the entirety of the United States, the UK, and Germany had been electrified. Large dams were erected and dozens of coal-fired power stations were built. Every household was connected to the grid and started to reap the benefits of lighting and many other inventions introduced in the early 1900s. The vacuum cleaner, washing machine, electric iron, electric oven, and home refrigerator, to name but a few, made life so much easier.

The development of the electric motor enabled a host of other inventions to be introduced in industry that led to many technical possibilities. Production costs dropped significantly as automated factory lines emerged in transportation, clothing, food, and many other sectors. This development created many icons of industrial society like General Electric.

Around the same time (1901) the first 'supergiant' oil field, Spindletop, was discovered in Texas. The field produced a huge quantity of 100,000 barrels each and every day. Crude oil became a rather cheap commodity. At first nobody knew what to do with such large quantities of oil, outside of its use as lamp fuel.

It was not long, however, before a number of inventions culminated in the emergence of the modern car based on the internal combustion engine. Even though it was invented in 1886 by the German Karl Benz, its low speeds and high production costs barred cars from widespread usage.

It was Henry Ford in the 1910s in the United States who, due to his keen understanding of the market combined with a prowess in technical innovations, developed the first multimillion version car sold in history, the famous Ford Model T. At mass production the car was four times as cheap as its closest competitor, the Oldsmobile, and essentially it could drive over twice as fast (65 km/hour) and could handle rural dirt roads with no difficulties. Car registrations soared from only 2.3% of households owning a car in 1900 to 90% in 1930.[44] The success was so large that previously common US rail transport was dwarfed entirely and has never recovered.

Today there are even fewer rail lines, fewer passenger kilometers, and much less goods transportation by rail in the United States than in the 1920s, despite the six-fold increase in the economy.[2] Not long after that, cars became a common sight in other more developed countries too, albeit at a much slower pace due to the First and Second World Wars. In the UK in 1951 only 14% of families had access to a car, but that rose to 60% by 1980.[3] And in Japan access to a car or truck grew from 1.2% in 1950 to 66% in 1970.[4]

44   Electric vehicles (EV's) in 2016 had a market share of only 0.08% of passenger vehicles, with Tesla having a 14% share of the electric vehicle market.

20.   What was the second chapter of the industrial energy revolution?

In OECD countries, after the end of the Second World War, dry-bulk container tankers fueled by oil were first introduced; in 1956 the first standardized containership based on an adapted oil tanker was sailing across the oceans. Shipping of oil itself in oil tankers had already become common decades earlier, and oil-fueled military ships had been introduced in the 1910s in the UK and United States. The first cellular containership was commissioned in 1960, and by 1982 1,000 such ships had been built which could handle 3,000 to 18,000 containers.[5] Today over 50,000 containerships are transporting millions of tons of goods across the globe.[6][45] The use of airplanes emerged just as rapidly as shipping. The first airlines using propeller planes for airmail and passenger transport were set up in Europe and the United States in the 1910s and 1920s. Transatlantic flights began in the 1930s, and the early much faster jet engine planes were introduced in the 1950s after the Second World War. The growth of air flight since has been so rapid that the size of the civilian aircraft fleet in 2015 reached nearly 25,000 planes.[7],[8]

Along with the development of transport and electricity, modern forms of heating were introduced, especially after the Second World War. Natural gas as the third fossil fuel after coal and oil had already been discovered in substantial quantities in Russia and the United States in

---

45   The switch from coal fuelled military vessels to oil based had been pushed by Winston Churchill in the UK as First Lord of the Admiralty in the 1910s, and simultaneously a similar switch was made by the U.S. navy.[84],[85] The innovation led to one of the decisive British advantages aiding the First World War victory, as UK boats could travel faster and carry greater weapon loads then those of the Germans.

the late nineteenth century. There was no way to use it, however, due to a lack of means to transport it, and most of the gas was simply vented in the air. The United States was the first to build gas trunk lines of 1,300+ kilometers in the late 1920s from Panhandle gas fields in Texas to Chicago and Detroit, followed by even longer trunks to New York and West Virginia.[9] After these pioneering efforts and new welding and pipe construction innovations, a pipeline construction boom started across the world. The former Soviet Union built the 800-kilometer-long Saratov–Moscow transmission pipeline in 1946, pioneered the construction of above-ground gasholder stations, and established the first underground gas storage station in 1957 to deal with seasonal demand swings.[10] Other countries also saw a gas pipeline boom following new large gas discoveries—including in the 1950s–1970s—in Germany, Hungary, Spain, the Netherlands (Slochteren), France, the UK, and Austria among others.

Natural gas quickly became a priority fuel for household heating and cooking in cities and densely populated countries. It also added to power generation in countries without abundant coal supplies or with depleted coal mines such as the UK. By 1975 natural gas supplied 10%+ of all energy usage in 10 out of the current group of 33 OECD countries, with proportions as high as 20%, 30%, and 50% in respectively Mexico, the United States, and the Netherlands.[11]

At the same time, the invention of the atomic bomb during the Second World War led to a race among the UK, the US, Canada, France, Japan, and the Soviet Union into nuclear power plant technology, with a wave of over 400 nuclear reactors built from the 1950s to the 1980s.[46] The

---

46   See for a world nuclear power plant database: www.world-nuclear.org/information-library/nuclear-fuel-cycle/nuclear-power-reactors/nuclear-power-reactors.aspx

expansion came to a standstill after the Three Mile Island accident in the United States in 1979 and the Chernobyl nuclear disaster in 1986 in the former Soviet Union, now at the border of modern-day Ukraine and Belarus, which spread radiation all over Europe. Even today no people live within a 19-mile radius of this former nuclear plant that is encased in a 400,000 m^3 concrete dome to prevent radiation leakage.[12]

It was a rude wake-up call for many who declared that an all-nuclear era was emerging. At the time it was anticipated in the majority of countries that nuclear would become the dominant source of power generation. President Richard Nixon in his 1973 address told the United States Congress that nuclear would provide more than 50% of power by the year 2000.[13] The only countries that have reached such a market share are France and Japan, before the disaster at Fukushima in 2011.

21. How did the 1970s-1980s oil crises change our energy outlook?

In the twentieth century, the large international oil companies from the United States and Western Europe (Shell, BP, etc.) —often referred to as the Seven Sisters— carved up most of the world's oil supply. Thanks to lucrative contracts made with leaders in oil-rich countries, crude oil was exported dirt cheap, for just a few dollars per barrel, with massive one-way benefits to consuming countries such as the United States, Germany, France, and the UK.

That all changed in the 1970s in the wake of a heated conflict launched by Egypt and Syria against Israel. Both wanted to recapture territories lost in the 1967 third Arab-Israeli war. To the surprise of many, Egypt decided to amass an invading army with Soviet support along the Suez Canal on the south and Syria on the north along the Golan Heights. It was a failure; Israel's stronger army pushed the forces back in a matter of days. A key part of the success was secret US support in supplying hundreds of tanks, artillery, and ammunition supplies via an airlift operation through Portugal code-named 'Nickel Grass.' In response to the US support, the Arab countries launched an oil embargo in the autumn of 1973, and halted all oil exports to Canada, Japan, the Netherlands, the UK, and the US. It was the start of the first oil crisis, which sent oil prices to about $60 per barrel (in 2014 dollars).

This oil crisis had such a substantial effect because US oil production had peaked in 1970, and the country could no longer increase its production levels. Earlier, in response to the 1967 Arab-Israeli war and US involvement, the Arab countries had tried an oil embargo and reducing oil supplies, but at that time the effect was negligible on world oil markets. Now, with real restrictions on available oil, rationing of gasoline became commonplace in the US until the embargo's end in March 1974, after the last of Israel's troops were withdrawn from west of the Suez Canal back to the 1967 Israeli borders, combined with intensive US diplomacy efforts.

Even though the embargo was lifted, the oil price barely dropped, as US production continued to decline while demand pressure continued. Not too long after this episode, the world experienced another oil shock, because of turmoil in Iran. The increasingly unpopular US- and UK-backed repressive monarchy of Shah Reza Pahlavi was overthrown in 1979 by a popular uprising inspired by a religious cleric who had been in exile since 1964, Ayatollah Khomeini. The steep reduction in Iranian oil output from 6 to 1.5 million barrels per day, as foreign workers fled the country, brought about an oil price spike towards $120 per barrel (in 2014 dollars), which dropped in the course of the 1980s.

The pressure of oil prices on economies and concerns over future security of supply in the 1970s-1980s led to drastically new energy policies, which have formed the basis for the clean energy revolution today. Budgets for clean energy sources were raised to $20 billion per year globally, and new technology avenues explored. While nuclear power received the majority of funding, a wide range of other sources also received a financial research injection, including solar panels, oil from algae, tar sands and shale oil, wind power, biofuels, and ocean thermal energy.

The research funding stream of the 1980s kick-started a lot of energy start-ups that worked on the continued development of solar panels and wind turbines. And while a lot of research money stopped after oil prices declined in the mid-1980s, Japan and the US pressed on with government spending on energy research. Also a number of corporations had seen the light, such as Vestas and Siemens, and they continued large research programs into clean energy. This proved fundamental for the continued development of solar panels and wind turbines.

Later on, in the course of the 1990s and 2000s, such programs received new impetus as a result of concerns about the impact on health of coal emissions and climate change caused largely by the incineration of fossil fuels.

In the past 30 years, the latest steps of innovation have led to modern silicon-based solar panels with 20% efficiency, steel tower wind turbines that operate in all wind conditions at 120 meters above the ground or higher, and all-electric cars with 300+ kilometer ranges.

The world is seeing a growing share of new renewables, with renewable sources in 2015 providing 26% of electricity supply. The transformation that is underway is driven by a renewables-based vision of the energy future started in Germany in the 1980s, the so-called 'Energiewende', and has been overtaken by China and the United States in this decade. All these revolutionary changes in the world of energy have brought us into a total new era. The Tesla Revolution has begun.[47]

47   The first electric passenger boats were already operating in 1894 but with the prominence of oil the story ended in 1926 until recently.

22.   When did we discover oil?

Crude oil has actually been known about since ancient times. In the seventh century, Byzantine soldiers used the substance to soak arrows in so that they could hail fire on the enemies of the empire. In ancient Mesopotamia, today's Iraq, oil was used to make ships watertight and for lighting. And Native Americans used it for baskets, weapons, and as medicine.

Despite this long history, large-scale extraction of oil started only 150 years ago, in the mid-nineteenth century, with the use of steam engines. These early versions operated by incinerating wood, coal, or other fuels to boil water that turns into steam, and the pressure of the steam creates mechanical movement of pistons for pumping or drilling motions. The first modern oil well was drilled with a steam engine, supervised by the self-styled 'Colonel', Edwin L. Drake. He developed his own drilling installation, which on 27 August 1859 in Oil Creek Valley, Pennsylvania, in the United States, reached a depth of 21 meters. This particular well provided a production of only about 1,500 liters per day, yet is seen as the birth of the modern oil industry.

The same technology was subsequently deployed at other sites after the news spread about Drake's drilling method. In the former Russian Empire, near Baku in the Caspian Sea—today Azerbaijan—the first steam engine-drilled oil wells were dug in 1873. The financing for the effort came from the Swedish brothers Ludwig and Robert Nobel, who later in life would discover dynamite and with their industrial wealth established the Nobel Prize for the sciences.

The Nobel brothers improved rudimentary distillation techniques to process crude oil into products, and built the first ship to transport large amounts of oil, an oil tanker.

The Baku site held so much easily extracted oil that at the turn of the century, around 1900, Russia was the largest oil producer with nearly 80 million barrels of oil per year. Today they produce 10 million barrels per day, the same as Saudi Arabia and the United States.

Not long afterwards, the development of other sites followed. The Yenangyuang oil field in Burma—today Myanmar—was put into production in 1889 by the British Burmah Oil Company, and the first oil well in India was dug in 1890 by Assam Oil Company. First exploration in Persia (Iran) followed when the mine owner William D'Arcy started exploration with financial support from the British Navy and Burmah Oil. The first commercial quantities of oil were struck in 1908, which led to the founding of the Anglo-Persian Oil Company, which later became British Petroleum (BP). Nearly all the truly massive oil fields were discovered several decades later, in the period 1935 to 1965, as oil exploration took off in the rest of the world and especially in the Middle East.

A distinction is made between fields that contain more than one billion barrels (giants) and more than five billion barrels (supergiants) of oil. Since 1965 the search for oil has only intensified, and while many oil fields are still found, on average each new field is a lot smaller. In the last 15 years, about 7,000 oil fields were discovered, yet their average size is about 26 million barrels, which seems large yet it is small in comparison to the 33 billion barrels of oil we use every year.[14],[15]

The giant and supergiant fields, mainly discovered in the middle years of the last century, still account for about 45% of total global conventional oil production.[16],[17] Production rates of number 2, 3, and 4 on the list of largest oil fields ever found are now declining. We are referring to

Cantarell in Mexico, Burgan in Kuwait, and the Daqing field in China, discovered in 1979, 1938, and 1959, respectively. The production of the largest oil field in the world, Ghawar, discovered in 1948 in Saudi Arabia, could in the near-term future begin its production decline.

23.   What was our main source of energy before oil?

Before we discovered large quantities of oil, coal was our main source of energy. The use of coal started in the sixteenth century in the United Kingdom where the first coal mine was opened. Thanks to this, the UK became the first nation to really start the first industrial revolution, towards a metals- and fossil-fuel based industrial economy.

The first use of coal was for home heating by burning, and for lighting by using town gas, which is a gas contained in coal seams formed alongside coal. Coal mining was first carried out manually by workers digging it out of the ground, which limited the speed of extraction especially since the depths of mining meant that coal mines would often flood and lay idle for weeks to months. This problem was overcome by the introduction of steam engines in the early eighteenth century that enabled continuous pumping of water out of the mines.

The coal mines using manual labor were slowly converted into mechanized enterprises in the late nineteenth century. The first coal-cutting machines were designed and built in the UK in the 1850s using compressed air to chip away at coal seams. The technology quickly spread to the United States and Russia, and by the turn of the century, nearly all mines employed such machines, with over 15,000 operating in the United States alone.[18],[19]

This development allowed for a fully mechanized supply chain including train transport powered by steam engines and fueled by coal. It was part of the parallel co-innovation of steel uses with coal, since coal was the ideal fuel to produce steel products of high purity and steel was used to build coal extraction and transport infrastructure.

Prior to the use of coal, large quantities of wood had to be used. Coal enabled new innovations to take hold in producing steel in the form of the blast furnace, where high-quality coal could be utilized to produce steel at scales not seen before. To the present day, no fuel works better than coal in blast furnaces for melting iron.

With the invention by Nikola Tesla of affordable electricity transportable over long distances, coal gained tremendous importance. Now it could be utilized to fuel a power plant wherever required. Rail lines could fuel power plants directly by transport from mine sites. The coal-electricity system was and is so cheap that practically overnight a lot of labor became redundant.

A kWh of electricity in the 1930s in the United Kingdom cost $0.31 (in 2014 US dollars), and a machine using this power could easily replace a full week of physical labour, at 43 dollars in wages.[20] Coal and electricity heralded the end of laborious work in the home and the beginning of production of goods on a 24/7 basis.

In the United States, the technological advances that utilized coal-electricity between 1900 and 1965 resulted in a reduction of 12 hours spent per week in preparing food, repairing clothes, and maintaining the house, which was instead spent on shopping, child care, and travel.[21] The ability to manufacture many goods in an automated fashion using an increasing diversity of mechanical machines contributed greatly to this, by enabling shaping and alteration of materials into desired products.

The combination of increasing electricity-powered mechanization and oil-fuelled transportation led to a transformation from an agricultural society where labor was mostly spent on food production plus construction with only a day of spare time a week, to a society where most labor was spent on manufacturing, with over two days per week and several hours during workdays to spend on family and leisure. Because of its low cost and energy-dense nature, coal is still used in the majority of economies in the world today. In 2015 coal was used to supply 37% of electricity and 28% of all energy used in the world.[22],[23] But due to the fact that coal is a major contributor to both smog and CO_2 emissions, the use of coal will decline in the next few decades, and thankfully can be replaced by sun and wind energy.

24.   What's the 'miracle of oil'?

The brilliance of oil as a fuel is its transportability, due to its liquid form, versus solid coal and gaseous natural gas. It also contains over 50% more chemical energy than coal, at a 'heating value' of 46 MJ of energy versus 26–33 MJ for the highest-quality coals.[48] Because of its liquid form, one super-tanker of two million barrels provides the same amount of energy as 30 full coal trains with 100 cars carrying 115 metric tons of coal each.[24] While natural gas contains even more energy, its gaseous nature made it difficult to transport in most countries, at least until the 1950s.

Crude oil also contains many different lengths of hydro-carbon chains that allow, after their separation in refineries via distillation and cracking, for a wide variety of products:

– **Kerosine**, the lightest fuel from crude is used mainly as jet fuel and for lamp oil.
– **Gasoline**, used for cars as well as aviation gasoline in light airplanes.
– **Diesel**, used for cars, is slightly heavier than gasoline and can be ignited just by compression, instead of requiring a spark like in a gasoline car.
– **Heavy fuel oils**, also called bunker fuels, are the heaviest of fuels from crude oil that are used primarily in shipping.
– **Naptha**, a large group of lighter carbon chain molecules used for the production of tens of thousands of chemicals used in paints, plastics, medicines, furniture and food products, among others

48   The physical reason is that oil contains a much more favourable carbon-to-hydrogen (C/H) ratio, at 7 to 1 versus a coal C/H ratio of 15 to 1. Because of the larger hydrogen content oil is much more energy-dense in nature.

- **Asphalt**, which is the heaviest part of crude oil and is used for roads.
- **BTX aromatics**, a special group of petrochemicals consisting of benzene, toluene, and xylene, used for hundreds of products including polystyrene packaging and polyurethane foams, like those found in synthetic pillows.
- **Refinery gas**, the light short carbon chain gas that comes out of the refinery distillation column, used to power the refinery itself.

The same products can be obtained from coal and natural gas, but that requires substantially more processing steps at higher cost. For example, in China many petrochemicals are currently produced from coal via methanol conversion. Methanol is a key chemical of the alcohol group already used to make various other chemicals used in plastics and products like antifreeze and plywood.

The qualities of oil are only half of the success story. Its entrenchment is also a story of entrepreneurship and capitalism intertwined with market distortion and political power plays.

At first, there were few uses for the fuel beyond lamp oil. It could not be used directly in steam engines and offered few advantages as a substitute source of fuel over cheap coal. As mentioned earlier, not long into the twentieth century, however, smart engineers had improved the design of automobiles to initiate the oil age.[2] And oil as a transport fuel for cars was only the beginning, with similar success stories for heavy-fuel oil-based shipping and planes powered by jet fuel since the 1950s.

In parallel, oil companies also found ways to utilize the light short hydrocarbon chains called naphtas. While these had been distilled in small quantities since the late nineteenth century from coal, wood, and oil, they had not found large use beyond lubricants and in rubber production. This all changed with the discovery of the first hard-setting plastic called Bakelite, in 1907, by the Belgian-American chemist Leo Baekeland. After commercialization in the 1910s, Bakelite was used for pipe stems, cigarette holders, telephones, and household appliances. Its durability and cheap cost made it the preferred choice for household materials.

Not long after, an surge of chemical research led to the commercialization in the 1920s to 1950s of the plastics we still use today, including polyvinyl chloride (PVC), Plexiglas, polyethylene (PE), polypropylene (PP), and polyurethane foam, among others.[25] Not to mention the tens of thousands of other chemicals from oil used for paints, detergents, soaps, coatings, colorants, pharmaceuticals, and explosives. It is said that over 50% of all items in a home or office have oil as their basis in one way or another.

A key political reason to substitute oil for coal for electricity generation was related to labor power. The liquid nature of crude oil makes it much easier to extract than coal and far less labor intensive, especially in the early decades of the oil industry. Moreover, oil is often extracted in other countries with dispersed supply chains instead of concentrated mines and railroads. Both factors mean that much less political power can be exerted by laborers to disrupt supply chains over issues of workers' rights.

The national coal strike in the UK in 1912 crippled the economy and led to ship fuel shortages, as one million coal miners laid down their tools and walked off the job. The government in response passed a minimum wage law to appease the strikers. Another general strike in the UK in

1926 was less successful, despite 1.7 million people of various trades laying down their tools to push for an increase in coal mine workers' pay.[49] The strike broke down after 10 days as the army rushed in to provide emergency supply, and the trade union congress of non-coal workers backed down.[26] The organized might of workers in the UK was a large influencing factor for politicians to favor oil over coal in shipping from the 1930s and later electricity generation in the 1950s. It allowed politicians to reduce the power of coal workers.[27]

49   In the United States coal workers at the time were paid three to four times as much salary per hour as in the United Kingdom.[20]

25.   Are we still discovering enough oil ?

Over the last 35 years, we have used more crude oil almost every year than we are discovering. During the 1960s, almost 556 billion barrels of oil were still being found (figure 5). This discovery rate has declined every decade since: about 395 billion barrels in the 1970s, about 193 billion barrels in the 1980s, and 125 billion barrels in the 1990s. Only since 2000 have discoveries stabilized at an average of 13 billion barrels every year.[15],[28]

Big Oil companies in Western countries have invested relatively small amounts of money in recent decades in the search for oil as there is not so much left to find. They preferred to spend their money on mergers and acquisitions, to enlarge their oil reserves, and on research and development to extract more oil from existing fields in their portfolios. Over the last few years especially, record low numbers of new oil fields were found, as exploration budgets were further hampered by the downturn in the oil price. In 2014 and 2015 only 8 and 3 billion barrels of oil were discovered, while production was 10 times as high.[15]

One of the few regions where large discoveries are occasionally made are deep-sea areas such as the Santos region in Brazil, where several large oil fields have been discovered up to depths of 7,000 meters since 2006. With up to 8 billion barrels of oil reserves each, these are exceptionally large finds. Most oil field discoveries today are a mere 26 million barrels. In the entire Santos area, potentially 50 to 100 billion barrels will be discovered.[36]

Giant and supergiant oil fields with multiple billion barrels are nowadays found every other year at best. The oil field Lula, discovered in 2006, was the largest discovery since 1999, when the Kashagan Field in Kazakhstan was

Fig. 6. Crude oil discoveries from 1960 to 2015.
Data source: [14],[15],[28]–[35]

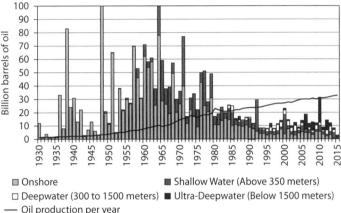

□ Onshore ■ Shallow Water (Above 350 meters)
□ Deepwater (300 to 1500 meters) ■ Ultra-Deepwater (Below 1500 meters)
— Oil production per year

discovered. Lula contains about 9 billion barrels of oil. In the golden age of oil exploration in the 1940s and 1950s, such finds were the rule and not the exception. A few more supergiant oil fields were discovered in Brazil after Lula with more than 5 billion barrels, including Sugar Loaf (Carioca) in 2008, and the Libra Field in 2010, in addition to a few smaller giants.[37]

Since we use 33 billion barrels of oil per year, we would need to find at least a handful of fields the size of Lula every year. Because this is not happening, we are eating into reserves from previously discovered fields. And the majority of new fields are found in difficult (expensive) locations, with 43% of newly found oil located at sea depths of 1,500 meters, and 16% between 300 and 1,500 meters deep.[29],[34]

To extract oil from fields like Lula, Petrobras first has to drill three kilometers to the sea bottom, then through a rock layer of two kilometers, and then through a salt layer of two kilometers before the oil is reached. Because of this type of complexity, production at the Lula field is only slowly ramping up, alongside several smaller giants in the Santos region, as development costs are extreme. In Brazil an oil price of $60+ per barrel is necessary to go to such ultra-deep-water depths.[38],[39][50] As oil prices in 2015/16 were far below this level, the majority of developments to bring deep-sea oil fields into production in Brazil have been postponed for at least several years.[40]

So we have been using more oil than we discover for a very long time. According to estimates by the United States Geological Survey, we have found over 90%–95% of all discoverable oil.[41]

---

50   Costs in Brazil are relatively low for ultra-deep-water due to the enormity of these fields and substantial existing infrastructure plus experience. In other countries a more typical cost value is 80+ dollars per barrel.

26. What is the history of solar energy?

Although the principle that light can be turned into electricity, the photoelectric effect, had been known since 1839—when the French scientist Edmond Becquerel stumbled on it in his experiments—the first solar cell was not built until 1941. It was Russell Ohl at Bell Labs in the United States who figured out that a junction across the silicon material could guide the electricity through it so that it could be tapped.

Because Ohl's first solar cell, which converted sunlight directly into electricity, had an energy efficiency of just 1%, it was an impractical device. Nowadays this efficiency rate is above 20% for cells sold in solar panels.

So this innovation lay dormant until a few years later when his colleagues were looking for a stand-alone way to produce electricity for the Bell telephone system, as batteries degraded too quickly in the Tropics. They substantially refined Ohl's junction design and in 1954 made the first practical solar panel at a 6% conversion efficiency. This triggered the interest of scientists at the US Space program, who would vigorously pursue solar advancement for space exploration.[42]

With practical commercial applications in place, many smaller entrepreneurs latched on to develop their own solar cells. Hoffman Electronics, a small radio and television firm, took the next big step. It rapidly learned how to manufacture solar cells, acquired a license for the Bell Labs solar cell patent, and by 1958 had an 8% efficient cell that was used in the first solar powered satellite (Vanguard I) to power its radios. Two years later, Hoffmann built a 14% efficient solar cell in the laboratory.[43]

Although further advances were made, it was not until the oil crisis of 1973 that solar power got the right attention.

From that point on, governments really started to speed up projects for alternative energy like wind and solar. The first tax credits for residential solar and wind were offered in 1978 in the US, and in the state of California the governor Jerry Brown initiated a $200 million solar program. The second oil shock in 1979, with oil prices skyrocketing because of a dollar crisis, advanced these efforts. President Jimmy Carter launched a $3 billion research program into solar electricity and heating.[42],[44] This can be seen as the real start of the current revolution in energy, which accelerated rapidly when oil prices shot up north of $100, around 2007.

The main user of solar energy in the early 1980s was the oil and gas industry, as it was ideal for powering offshore oil rigs. About 70% of all solar modules sold in the US (about 80% of the global market) were used by oil and gas companies. Because of these developments, oil producers Exxon, Mobil, ARCO, and Amoco bought solar manufacturing companies in the 1970s and 1980s, which they sold again in the mid-1980s to 1990s. Notably ARCO, a US oil company bought by BP in 2000, had the largest solar manufacturing capacity globally in 1980 at 1 MW per year (today all factories together produce 50,000+ MW annually). The solar unit of ARCO was sold by BP in 1990 as it never generated any profits. The buying company Siemens in Germany a few years later tried to sue ARCO, as it alleged it was unaware of the lack of profitability of its solar business at the time of purchase.[43]

The cost of solar energy at that time had dropped significantly to about $30 per watt from $80 five years earlier. The future for solar looked bright, but it came to a standstill in the United States because the solar program ended as president Jimmy Carter was replaced by Republican Ronald Reagan. Solar panels were removed from the White House, research spending was slashed, and solar tax credits came to an end by 1985. The only key source of funding now was the oil industry which invested in their own solar companies, blocking new entrants from entering the market.

Fortunately, solar research did not end there, as in parallel, Siemens in Germany and Sharp, Kyocera, and Sanyo in Japan had made their own moves into solar, best known for their solar powered calculators.[43] The setup of large manufacturing facilities in the 1980s on both continents got a further boost from government support programs. The German government in 1990 launched a 1,000-roof solar-PV program and instituted the first feed-in electricity tariff in 1991. It required utilities to connect solar energy installations to their grid systems at 80% of the historical retail price. Siemens Solar, which had previously purchased ARCO's solar unit, installed 20% of total global solar-PV capacity in 1996, at 20 Megawatts.[43] In the wake of Germany's policy, the Japanese launched the 10,000-solar roofs program, and instituted its own feed-in tariff that obliged electric utilities to purchase power at the retail price. Solar efficiency further improved, and Kyocera in 1993 achieved a laboratory-scale efficiency of 19.5% in converting sunlight. Quickly Japan overtook all other countries, accounted for 40% of world solar-PV production by 2000, and installed 50,000 solar systems on rooftops.[45]

A new dawn for solar began as 1,000 MW were now installed. With industry maturation and massive innovations, solar costs had dropped from about $80 per watt in the early 1980s, to about $5 at the turn of the century. By 2007 over 37 countries had set up feed-in tariffs to support solar and other renewable energies.[46] US president George W. Bush symbolically reinstalled a 9 kW solar panel system on the White House lawn.[47] BP and Shell both ventured into solar panel manufacturing (to exit again five years later).

Since then, the key change has been market consolidation in China as the leading solar panel manufacturer in the world, with over 90% of the supplier market. The key reason has been a race to the cheapest solar energy cell pushed by the Chinese government. Already in its 10th Five-Year plan (2001–2005) the country pushed solar research and development to a goal of 5 MW of annual production capacity, followed by the 'the Renewable energy law' in 2006 as part of the 11th Five-Year plan, which obliges power companies to purchase and dispatch solar and wind electricity. The slump in investment in Europe in 2007/08 following the financial crisis helped China to take over manufacturing firms and acquire solar technologies, with over 124 acquisitions taking place between 2002 and 2012.

China quickly became the leading solar exporter. Its own solar panel market was still small, and as it outcompeted with low prices, solar manufacturing in Germany, the US, and Japan virtually disappeared. Large cost-cutting exercises, cheap Chinese labor, and hundreds of innovations made the price of panels drop further to below $1 per watt in the year 2012, while the efficiency of commercially available panels grew to 15%-20%, with over 40% efficiency reached at laboratory scale. In many ways Germany's favorable feed-in policies helped the Chinese solar industry to mature, with every German installed solar panel being 'Made in China.' Only when China instituted a $0.15 solar tariff in 2011 as part of its 12th Five-Year plan did solar panel sales take off rapidly in China itself. Today China is #1 in solar installations, followed by Germany, Japan, and the United States. Solar panel installations globally have grown to a whopping 100 gigawatts in 2012 and 230 gigawatts in 2015.

27.   How much does it cost to live off the grid?

Living off the grid, with your electricity coming from just solar energy, is increasingly becoming a reality. Since 2010, the cost of solar panels has come down 50%, with a 5 kW cost of $7,300–$8,500 in 2016. This is sufficient to cover all daily use for larger households in most high-income countries.[48]–[51]

The system cost excluding any subsidies translates to $0.09 per kWh in cloudy countries like Germany and $0.05 per kWh in sunnier countries such as Australia (all costs divided by total generated electricity during the life of the solar panels).[51] So solar power is now less expensive than the incumbent grid costs for most households, even when excluding taxation and levies (except when households need to borrow money for their system and pay interest).

But the solar energy revolution is not complete yet, since to go truly off the grid requires a storage system for the low sun hours in the evening and at night. Currently the leading choices are either lead-acid or lithium-ion batteries. The size of such a system varies by the solar radiation per hour in relation to power use during the day, but as a rule of thumb, 60% of electricity in a household is used in low-sun to no-sun hours.[52] In the case of Sydney, Australia, where there is sun all year round with only a minor slump in the

---

51   Calculation based on a lifetime of 25 years for 240 Watt peak solar-panels with a 1620x986 mm dimensions, a solar radiation of 1100 kWh/m2/year for cold countries and 1900 kWh/m2/year for warm countries, a solar system efficiency of 15%, a systems efficiency rate of 80%, a packing factor of 0.94, and a solar output degradation rate of 0.7% based on experience with 2000 solar systems.[86],[87] The values also excludes any subsidies.

52   Rule-of-thumb based on minute by minute simulation of domestic electricity demand using the CREST model developed by Ian Richardson and Murray Thomson of Lougbourough University.

Fig. 7. Solar generation per hour in kWh (top) and electricity demand on the bottom met by solar power directly (grey), battery stored solar (black), and grid electricity (white), for 14 days early in July (Winter) in Sydney, Australia.

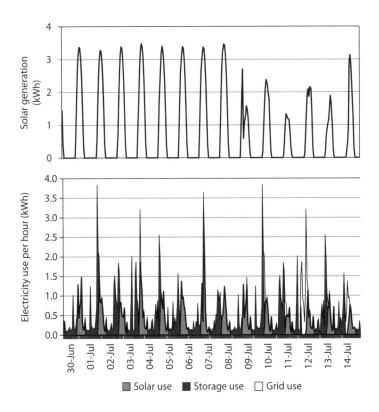

autumn/winter months of April to August, about 9.5 kWh per day or one Tesla Powerwall 2.0 is needed to cover normal electricity demand during the night (figure 7).[53] Even in such a sunny country, there are some days with limited sunshine.

53 Based on average Australia household electricity use, hour by hour solar generation profiles generated using Renewables Ninja (www.renewables.ninja).

On these days either much less electricity must be used or a grid backup is needed to fill the gap.

The total solar installation plus Powerwall system would bring costs up to about $0.16 per kWh,[54] below the 2015 electricity price of $0.22 per kWh.[52] Living off the grid in Australia and other sunny countries and regions, with a minor inconvenience of not enough power for a few hours every month, is within reach for households, using today's technology.

There is still one big caveat for countries with dark winters like Germany where solar with battery systems do not generate sufficient electricity for several months. For example, in Berlin, Germany, on some winter days in November through January, hardly any solar power can be generated. Large-scale storage or a backup grid connection is necessary to bridge this seasonal problem. Grid backup is the choice for an increasing number of German households who are living off the grid at least part of the year. In Germany by mid-2016 about 40,000 households and small enterprises had installed power storage units together with their solar installations. Battery sales are exploding in Germany with nearly half of all new solar installations equipped with backup batteries and a $1 billion battery market forecast by 2021.[53]–[55][55]

---

54 The calculation includes an assumed power-wall replacement after 10 years with a new system at 50% the present costs.

55 Their popularity is thanks to low cost systems by Tesla's battery competitors, such as the German company Sonnen (which stands for Sunshine in German). Its latest eco-4 lithium-iron-phosphate battery with a 4 kWh capacity that retails for $6000 excluding installation costs.[88] Based on no less than 10,000 cycles warranty and assuming an 8 kWh system, total costs for a solar + storage system over 25 years amount to $0.28 per kWh today in Germany, below household electricity prices (assuming no loans).

The German developments in off-grid electricity have led to horror for electricity providers who have coined going off the grid 'grid defection' since it destroys their business model.[56] With declining solar-PV and battery costs, utility providers worldwide will have to adapt, as off-grid living will become increasingly popular in the next decade. We are already seeing this happening for the 1.2 billion people who still have no access to the electricity grid today, and another 1 billion with an unreliable service, typically backed up with costly diesel power. In these markets, the biggest success to date is a combination of pico-solar solutions, up to only 10 watts, combined with super-efficient appliances and pay-as-you-go financing. This makes it far more afford-able to get solar with small batteries than other alternatives like kerosene lamps for lighting or diesel generators for electricity.

It has been estimated that between 2011 and 2015, over 100 providers entered the pico-solar market, with sales of 44 million solar systems smaller than 10 watts.[57] The systems come equipped with a microbattery that after sunset enables eight hours of lighting, mobile phone charg-ing, and can also power super-efficient radio, fans, and a small television. Since these types of households don't have access to traditional banking services, a system has been developed where it can be leased using mobile phone money wallet systems, such as the 20-million-users M-Pesa sys-tem launched in 2007 by Safaricom.[58] One of the largest pico-solar providers—M-Kopa in Kenya—has so far leased 400,000 of their 8 watt systems to households at a cost of $0.50 cents per day for a year, with sales of 1 million units expected in 2017.[59],[57]

For entire communities, solutions in the form of solar-based microgrids are now available. They range from 0.8

kW to several megawatts with a battery storage unit and local AC electricity grid lines. The biggest development is anticipated in India, where the government has announced that solar microgrid services should be rolled out to 25,000 villages in India.[60] The private sector's response has been somewhat muted. The largest initiative to date has been undertaken by Boond Engineering Company which provides a system with a 0.8 to 2 kW set of solar panels and a battery with distribution system for up to 50 houses, at a cost of up to $2 per month per family.[61],[62]

28. Is wind energy a great success as well?

Large amounts of financial support for wind energy have created the image that wind turbines don't run on wind, they run on subsidies.[63]–[65] Many argue that government support for renewables is a waste of money that is better spent by taxpayers themselves.

Even though prototypes of modern windmills were built in the 1940s and 1950s in the US and UK, the real start of wind energy had its origins in Denmark. The Danish machine and agricultural companies Vestas and Danregn Vindkraft started constructing steel wind turbines with fiberglass blades at 10–15 meters height in 1980. Larger versions of their wind turbines were also exported to the United States supported by US tax credits following the second oil shock. In a difficult transition process, Vestas, on the brink of bankruptcy in 1986, decided to focus purely on the construction of wind turbines. In hindsight it was a highly successful move, and the company is now the second-largest wind turbine manufacturer in the world after Goldwind in China.

By 1990, over 1,000 wind turbines had been sold by Vestas worldwide, and in 1995, they sold the first 600 kW version with a height up to 60 meters and 22-meter blades. It took only four more years to get to the 2 MW turbines, which are the standard size for onshore wind turbines today.[66] The cost of wind turbines dropped rapidly in this period.[67][56]

---

56 The innovations in wind energy was also helped thanks to Enercon, the German wind giant, established in 1984, Zond in the US, which stepped into manufacturing in 1993 (now General Electric Wind), the wind manufacturers Gamesa (now a part of Siemens) plus Suzlon Energy formed in Spain and India in 1995, and Siemens which purchased Danregn Vindkraft (then called Bonus Energy) in 2004.[66],[89]

Since its origins in the 1980s, modern wind power has only been financially possible with market support, either from direct government subsidies or a feed-in market surcharge on the electricity price fed back to wind investments. The challenge with wind energy has been a stabilization in cost declines, and even cost increases from 2000 to 2009, due to lower innovations and all-time high steel and copper prices. A modern 2.6 MW wind turbine uses about 275 tons of steel and 4 tons of copper.[68]–[70]

The absence of cost declines plus continued government support has fueled the view that wind will never become commercially viable. In the top three wind energy countries, with in total 60% of all onshore produced wind energy in 2015, the following financial support systems are in place:

–   **China** with an installed 145 GW of capacity has become world leader in wind energy in less than 10 years since its renewable energy law was launched in 2006, quickly followed by a feed-in tariff in 2009 at around $0.09 per kWh onshore wind support.[71],[72] The support will be reduced by 2018 to $0.03 to $0.05 per kWh, varying by region.[73]

–   In **the United States** with 75 GW wind power installed, a Production Tax Credit (PTC) has been in place since 1992, providing $0.23 per kWh to wind farms for a 10-year period. The government is phasing out these credits with a 60% reduction by 2019 and a complete phase-out by 2021.[74] Additionally, individual US states such as Texas and California have put their own support systems in place.[57]

---

57   In Texas one of the legacies of then governor George W.Bush, has been a massive expansion in wind energy capacity, as he signed to law in 1999 mandated wind energy capacity levels combined with a support auction system.

- The installed 45 GW capacity in **Germany** has been backed, since 1991, by a feed-in tariff which in the last few years has been held around $0.10 per kWh.[75] As of January 2017, the system will be replaced by an auction system.[76]

Since 2009, costs have started to decline again, thanks to a drop in steel and copper prices of more than 50%, a shift from steel to concrete towers for wind turbines, much larger wind turbines of up to 5-6 MW capacity, and lower interest rates.[77],[78] So less financial support is needed to make onshore wind farms happen. In 2016 a new record was set with an 850 MW auction in wind-rich Morocco. The winner was a $0.03 per kWh offer by Italy's Enel Green Power, supported by Siemens' wind turbine division.[79] Previously, in Canada's Quebec province, a 450 MW onshore park was awarded with a bid price of almost $0.06 per kWh including transmission costs.[80] Important to note is that the support here is a price and sales guarantee, not a subsidy. Similarly, unexpected record lows per kWh were set for offshore wind in 2015 and 2016 in tenders in Denmark and the Netherlands, for 400 and 350 MW parks, at a price of $0.11 and $0.08 per kWh.[81],[82][58] An even lower price of $0.07 per kWh was tendered later in the year for a 350 MW wind park near the Danish shore.[83]

---

58   Vattenfall and Dong Energy came with the best bids. The Dong energy project when including transmission costs, provided at no-charge in the tender, would come in at 9.6 USD cents per kWh.[90]

Chapter 3 – The Petrodollar and the Geopolitics of Oil

If we had not sent troops into Jordan, not only would there have been violence but we would have lost our oil. We have spent large sums of money in the Middle East, and all Russia has paid into that area is propaganda. British foreign policy should never be afraid to say we are there to protect British interests.
– Margaret Thatcher on the British military intervention in Jordan, 1958

It is not difficult to understand. The most important difference between North Korea and Iraq is that economically speaking we had no choice with Iraq. The land is swimming in oil.
– Paul Wolfowitz, former US secretary of defense,1999

People say we're not fighting for oil. Of course we are. They talk about America's national interest. What the hell do you think they're talking about? We're not there for figs.
– Chuck Hagel, former US Republican senator and secretary of defense in the Obama administration, 2007

Introduction

The early history of industrial society consisted of a bloody struggle to carve up the world among a handful of European countries: the UK, the Netherlands, France, Portugal, and Spain. In the early sixteenth century, these powers dominated the world thanks to early advances in ships, gunpowder, and mechanized military. At the time of the First World War (1914–1918) they had colonized nearly all of Africa and South America and large parts of Asia. As a consequence, an enormous gap in wealth emerged between rich and poor countries, which still largely exists today.

The struggle for dominance and political influence became intertwined with oil from the First World War onwards. Oil has been, at least for the last 100 years, the greatest prize, and its geopolitical importance as a strategic energy commodity cannot be overstated. Especially the two Gulf Wars have demonstrated that the United States will use any crisis or opportunity to deepen its influence in the Middle East. Dozens of such wars have been fought throughout history over the prospect of accessing or controlling oil supplies, in combination with a strategy of shrewd deals with local elites or the installation of puppet governments by covert means.

The history of oil as a geopolitical instrument can be understood best from the global military and political interventions of the United States, the United Kingdom, and Russia/the Soviet Union. Especially the UK and, later on in history, the United States, benefitted hugely from these practices by accessing the mineral and oil wealth of many countries while providing little in return. Only since the end of the colonial era in the 1950s–1980s has there been a slow change in the geopolitical balance, as most countries are now independent, industrializing, and gaining economic and political power.

29.   Oil and power—who is leading whom?

In the early age of oil, about 150 years ago, oil companies in the United States and Europe sought to control oil supplies, mainly in pioneering regions of their home countries. The key was to gain possession of refining and distribution networks as these provided the ability to set prices as well as market power. The biggest oil firm ever in history by far, Standard Oil of the United States, had captured 90% of the refining market by 1900, while production was only 0.4 million barrels per day.[1]

Once the twentieth century started and more and more oil discoveries were made overseas, US, UK, French, and German oil companies, together with their governments, began to use their full power to control the entire supply chain of oil.

In the period 1900–1915 the first major oil discoveries in the Middle East were made in Persia (today's Iran). In *A Century Of War: Anglo-American Oil Politics and the New World Order,* William Engdahl describes how the British soon understood its importance and used the First World War to put the recently discovered oil fields in the Middle East under British mandate.[2] This was arranged through the secret Sykes–Picot Agreement in 1916 between the British and French. It divided the Arab provinces of the Ottoman Empire (today Turkey) outside the Arabian Peninsula into areas of future British and French control or influence and also aimed to create an international independent zone in Palestine.

Soon after the Sykes–Picot Agreement, in 1917, the Balfour Declaration was published by the UK's Foreign Office. It is widely regarded as the start of the new state of Israel. The declaration stated:

His Majesty's government view with favor the establishment in Palestine of a national home for the Jewish people, and will use their best endeavors to facilitate the achievement of this object.

The declaration was later incorporated into the Treaty of Sèvres of 1920 that divided up the lands of the Ottoman Empire among France, Britain, Armenia, and Greece. The territories gained by Britain and France was largely in line with the earlier Sykes-Picot Agreement.

This turn of geopolitical events stood in sharp contrast to the so-called McMahon-Hussein correspondence, established earlier in 1916 between Britain and the Emir of Mecca, also the self-declared king of Arab lands under Ottoman rule. In this correspondence, control over the Middle East territories was promised to the Arab independence movement so they could create an Arab nation 'in the limits and boundaries proposed by the Sharif of Mecca' in exchange for their revolt against the Ottoman Empire during the First World War, which took place in June 1916.

The issuance of the Treaty of Sèvres initiated geopolitics over oil in the Middle East. The background of British support for a Jewish homeland in Palestine at the time was linked to geopolitical calculations. After 1920, Palestine was occupied under a British mandate with a military force of 100,000. The influential geographer and cofounder of the London School of Economics, Halford J. Mackinder, expressed his view that the occupation of Palestine would strengthen Britain's position on the Suez Canal, reinforce the route to Great Britain's imperial dominion in India, and provide an important buffer to Egypt as it geographically blocked it off from today's Syria, Jordan, and Saudi Arabia.[2][59]

59   The U.S. had been involved in the oil industry in the region since 1922.

30.   What happened after the Second World War?

By the 1930s, the United States had joined European countries in the hunt for oil, as Standard Oil of California (SoCal) gained large oil concessions in 1933 from King Ibn-Al Saud, who had named his new kingdom after himself: Saudi Arabia. After striking sizeable quantities of oil in the Rub-Al-Khali desert in 1938, the United States shifted its position, and Saudi Arabia became a key geopolitical ally.[3]

Not long after the end of the Second World War, the United States began to expand its efforts on economic, military, and political fronts to gain further geopolitical control in the Middle East, largely focused on stability and the flow of oil, not just from Saudi Arabia but also the other major states around the Persian Gulf (Iran, Iraq, and Kuwait) given that they were major producers of oil and also surrounded the oil tanker flow in the Persian Gulf. In the decades from the 1950s to the 2000s, over a dozen government coups have been orchestrated through US intervention.[4]

The early architects behind these coups d'état, or even US-executed political assassinations, were the infamous Dulles brothers. In the Eisenhower administration, these two brothers had risen to the posts of secretary of state and director of the CIA, from which they exerted their influence between 1953 and 1959. After the democratically elected reformer Mossadegh nationalized the oil industry in Iran in 1953, the Dulles brothers succeeded in removing him, after which the shah, an American ally, was made the new ruler of Iran. They had done so by large intelligence infiltration, fueling existing anti-Mossadegh insurgency, and bribing Iranian military officers to take up arms again Mossadegh.[4],[5]

Similarly, the CIA had a hand in the coup in Iraq in 1963 to overthrow the anti-American government of Abdel Karim Kassem, who himself five years earlier had taken over power from a Western-allied government. This coup was carried out by a new Arab group in the Middle East, the Baathist Party, which included none other than a young Saddam Hussein who would later rise to power as Iraq's dictator. According to Roger Morris, former member of the US National Security Council in the 1960s, the coup rapidly opened up oil activities in Iraq to US and British companies:

> The United States also sent arms to the new regime, weapons later used against the same Kurdish insurgents the United States had backed against Kassem and then abandoned. Soon, Western corporations like Mobil, Bechtel, and British Petroleum were doing business with Baghdad.[6]

Another part of the US strategy was the provision of substantial support to the new state of Israel as the only Western-oriented non-Arab country in the region. Israel was officially formed under international and domestic pressure after Britain ended its Palestinian occupation following the Second World War. After the 1948/49 Arab-Israeli war, it managed to secure its new borders. Later, as Soviet influence in the Middle East grew, and in response to an unsuccessful attempt to remove the Hashemite leadership in Jordan by the Palestine Liberation Organization, the US as of 1971 began to provide $2 billion per year to Israel, of which two-thirds in weapons.[7]

Noam Chomsky has several times pointed out the fact that a strong Israeli military is in the interests of the United States:

Since 1967 it's been plain that Israel is, by a long shot, the strongest military power in the region [...] American planners have regarded Israel as a barrier to Russian penetration [...] Israeli power protected the 'monarchical regimes' of Jordan and Saudi Arabia from 'a militarily strong Egypt' in the 1960s, thus securing American interests in the major oil-producing regions. For such reasons, the United States has tacitly supported the Israeli occupation of surrounding Arab territories.

As explained earlier, the United States and European involvement in the Yom Kippur War in 1973 led to an oil embargo against Canada, Japan, the Netherlands, the United Kingdom, and the United States by OPEC.[60] It demonstrated Saudi Arabia's power as the largest oil exporter in the world. Global oil supply dropped by 5% within a year, and the price of oil shot up from $18 per barrel to nearly $55, an oil shock with many long-lasting geopolitical and economic effects.[8][61] Since oil is traded in US dollars, any oil price increase automatically results in an increase in the demand for dollars. This greatly helped the United States in fighting declining confidence in the dollar that had started in 1971 when President Nixon ended the gold standard.[62]

The United States was so 'distraught by being challenged by underdeveloped countries', in the words of then defense secretary James Schlesinger, that they considered military action to seize Middle Eastern oil fields in 1973. The United States, which at the time imported over 10% of its oil from

60   In 1960 the Organization of the petroleum exporting Countries (OPEC) was established by Iran, Saudi-Arabia, Kuwait, Iraq, and Venezuela. At the time they delivered 85% of all exported crude oil to the European and Japanese market.
61   Values in inflation adjusted 2015 dollars.
62   Prior to 1971 the US dollar was backed by gold.

the Middle East[63], was of the opinion that they could not 'tolerate a situation in which the United States and its allies were at the mercy of a small group of unreasonable countries.' Secret plans for a military operation to seize oil fields in the eastern strip of Saudi Arabia, Kuwait, and Abu Dhabi were formed, and had it not been for the end of the oil embargo in March 1974, a two-brigade-strong military force with air force backup would have landed, according to declassified British documents from the United States to the British prime minister.[9]

After the embargo, several countries initiated policies to contain their dependence on oil. It led to much greater interest in renewable energy and nuclear power. Brazil implemented its Proálcool (pro-alcohol) policy in 1975 mandating mixing of ethanol with gasoline for automotive fuel, which is still used to this day. The crisis also led to a shift in the Japanese economy away from oil-intensive industries to sectors such as electronics, and in Europe it greatly reduced the demand for large cars. To prevent any short-term disruptions, high-income countries set up strategic oil reserves from which oil could be supplied for at least a few months, governed by the International Energy Agency set up in 1973.

The geopolitical competition between the Soviet Union and the United States gave Arab states 'disproportionate international bargaining power.'[55] In 1970 the United States helped Anwar Sadat, an activist in the Muslim Brotherhood, to secure his presidency in Egypt after the sudden death of Gamal Abdel Nasser. For two decades the US had covertly supported the Muslim Brotherhood, as they saw Nasser as a threat to US interests because of his calls for pan-Arab

63   Compared with 80% for the Europeans and over 90% for Japan.

unity, anti-Israel rhetoric, secret weapon dealings with the Soviet Union, and nationalization of the Suez Canal from British and French interests.[10] Sadat reoriented his country from Russia to the United States and as of the 1979 Israel-Egypt peace treaty received billions of US dollars in aid and weapons.[11]

After the 1979 Soviet invasion of Afghanistan, several states in the region asked the United States for security guarantees against such Soviet military aggression. Since then, the US and UK have been the prime arms dealers to Arab nations, as part of a deliberate strategy to sell arms to allied states in the Middle East in exchange for dollars made from oil sales.[12]

Saudi Arabian arms purchases from the US started in the 1970s with over $6 billion spent, and accelerated in the 1980s after the 1979 Iranian revolution. A few of the deals made by the UK and US with the Saudis include:

– The US sold five special AWACS radar planes and eight refueling aircraft in 1982 for $8.5 billion to the Saudis.[13]
– The UK prime minister Margaret Thatcher in 1985 initiated the Al-Yamamah arms deal with Saudi Prince Bandar for the sale of over 100 fighter jets costing $43 billion, paid for by a 20-year delivery of 600,000 barrels of oil per day.[14],[15]
– And as recently as 2010, the US made a deal for $60 billion in weapons sales to Saudi Arabia.[16]

To this day the rivalry between Saudi Arabia and Iran continues. In 2010, WikiLeaks published diplomatic cables that revealed that the Saudi King Abdullah had urged the United States to attack Iran in order to destroy its potential nuclear weapons program.[17]

31.   Wasn't the war in Iraq after 9/11 all about oil as well?

Iraq, after Saudi Arabia, has the largest cheap oil reserves in the world, and most importantly, its fields are still largely undeveloped. Former US vice president Dick Cheney, who led the White House into the Iraq war in 2003 under the G. W. Bush government, told an audience of oil analysts in London in 1999 that the biggest prize could be found in the Middle East, given the rapidly shrinking abundance of new oil prospects.[18] Cheney also happened to be a strong supporter of the interventionist Dulles 'ethic', to preemptively topple foreign powers to the economic benefit of the United States. He shared such views with his former close colleague of 30 years, Donald Rumsfeld, whom he successfully put forward to G. W. Bush for the post of US secretary of defense.

The twin tower attacks of 9/11 presented the perfect event, for Cheney and Rumsfeld, to unleash a strategy of preemptive wars. Leaked memos reveal that just over a week after 9/11, the Pentagon's Defense Policy Board held a 19-hour meeting on 19/11, discussing how to overthrow the Iraqi government of Saddam Hussein. The meeting was chaired by Rumsfeld and deputy-Secretary of Defense Wolfowitz, and among the 18 participants were Henry Kissinger, CIA director James Woolsey, James Schlesinger, and Richard Perle of the Project for the New American Century.[19],[20] The US general Wesley Clark confirmed this during a promotion tour for a potential presidential bid. He explained why the neoconservatives in the G. W. Bush administration had decided to 'take out seven countries [...]: Iraq, Syria, Lebanon, Libya, Somalia, Sudan, and Iran.'[21] All of these had been longtime friends with

Russia, and the US wanted to bring them under American control;

The truth is, about the Middle East is, had there been no oil there, it would be like Africa. Nobody is threatening to intervene in Africa [...] There's no question that the presence of petroleum throughout the region has sparked Great Power involvement [...]there's always been this attitude that somehow we could intervene and use force in the region.

About 10 days after 9/11, I went through the Pentagon and I saw Secretary Rumsfeld and Deputy Secretary Wolfowitz. I went downstairs just to say hello to some of the people on the Joint Staff who used to work for me, and one of the generals called me in. He said, 'Sir, you've got to come in and talk to me a second.' I said, 'Well, you're too busy.' He said, 'No, no.' He says, 'We've made the decision we're going to war with Iraq.' This was on or about the 20th of September. I said, 'We're going to war with Iraq? Why?' He said, 'I don't know.' He said, 'I guess they don't know what else to do.' So I said, 'Well, did they find some information connecting Saddam to al-Qaeda?' He said, 'No, no', he says, 'There's nothing new that way. They just made the decision to go to war with Iraq [...]'.

I came back to see him a few weeks later, and by that time we were bombing in Afghanistan. I said, 'Are we still going to war with Iraq?' And he said, 'Oh, it's worse than that.' He reached over on his desk. He picked up a piece of paper. And he said, 'I just got this down from upstairs'— meaning the Secretary of Defense's office — 'today.' And he said, 'This is a memo that describes how we're going to take out seven countries in five years, starting with Iraq, and then Syria, Lebanon, Libya, Somalia, Sudan and, finishing off, Iran.'

It took Cheney's team a year to convince enough US senators and Tony Blair's government in the UK to vote for the invasion on 11 October 2002, largely on the basis of falsified and massaged intelligence from the CIA about the possession of weapons of mass destruction by the Iraqi dictator Saddam Hussein, as there was no 'smoking gun.'[22] This was admitted later by Colin Powell, US secretary of state in the Bush-Cheney government and a widely respected retired four-star general, who gave the most convincing speech to invade Iraq at the behest of Cheney.[23]

In his 2005 book *Petrodollar Warfare*, William R. Clarke explains that the US-UK decision to invade Iraq in 2003 was oil driven.[24] According to him, the petrodollar system was the driving force of US foreign policy. It seems to be no coincidence that the Bush family has had close personal ties with the Saudi royal family since the 1970s.[25] And even Alan Greenspan, who served as chairman of the US Federal Reserve for almost two decades, wrote in his memoirs:

> I am saddened that it is politically inconvenient to acknowledge what everyone knows: the Iraq war is largely about oil.

The strategy of preemptive war was described as early as 2000, surprisingly one year before 9/11, in a detailed report called 'Rebuilding America's Defense' from the Project for a New American Century (PNAC).[26] Ten founders of this neoconservative think-tank were members of the George W. Bush Administration, including Cheney, Rumsfeld, and Wolfowitz. One of the report's neoconservative writers, Robert Kagan, revealed in an interview with the *Atlanta Journal Constitution* in 2002, 'We shall need a large armed force for a long period of time in the Middle

East. If we have a military force in Iraq, there will not be any disruption in oil supplies.'[27] The PNAC-report even stated new strategies, like preemptive wars would never be acceptable to the outside world without a 'new Pearl Harbor'.

32. Didn't oil play a role in the first Gulf war too?

In his classic *A Century of War,* author William Engdahl wrote that the US had searched for years for a good excuse to station a military force in the Middle East at the end of the 1980s.[2] At the end of July 1990, relations had become tense over oil drilling by Kuwait into the Iraqi Rumaili oil field shared across their borders. Iraq's dictator Saddam Hussein claimed that Kuwait had stolen $2.4 billion from Iraq in this way, and that Kuwait had depressed the oil price by excessive production, thus starving it of income. There was open talk of war as Hussein threatened that 'we cannot tolerate this type of economic warfare.'[28]

The American ambassador in Baghdad, April Glaspie, asked for a meeting with Saddam Hussein in the midst of the heated situation. The conversation of 25 July 1990 indicated that, according to an Iraqi transcript, Washington would not interfere if Iraqi military force were used. Glaspie, according to the transcript, had said, 'We have no opinion of the Arab-Arab conflicts, like your border disagreement with Kuwait.' Thus instead of a strongly-worded warning against any armed action, she had given tacit approval to invade Kuwait.[29]

This was the signal that Saddam had been waiting for, and his troops invaded eight days after the meeting with Glaspie, conquering Kuwait within a week. The ruling royal al-Sabah family had fled shortly before with all their gold and jewelry, probably tipped off by the CIA.

This interpretation has been confirmed by the Russian diplomat Primakov, according to Engdahl. He had traveled to Washington and London in October 1990 shortly before 'operation Desert Storm', the first invasion into Iraq's occupied Kuwaiti territories, with a proposal from Saddam

Hussein by which 'a war could have been prevented.'[2] Yet both US president Bush Sr. and the UK prime minister Margaret Thatcher were not amused by the news.

Iraq's occupation of Kuwait began on 2 August 1990. Soon after, the US deployed military forces in Saudi Arabia, as agreed with the Saudi King Fahd, and it urged other countries to do the same.[30] By November 1990, a 34-country coalition was formed under the flag of approved use of force by the United Nations security council, led by Saudi, US, and British forces, joined by many Arab and Western countries.[31] The military campaign to expel Iraqi troops from Kuwait started after US Congress approval on 12 January 1991, with an aerial and naval bombardment on 17 January followed by a ground assault on 24 February. Coalition troops gained a large victory within four days, and Iraq was pushed back into its own borders.[32] US President Bush Sr. after this success decided not to advance into Iraqi territory, and the coalition troops never reached Baghdad, a decision that surprisingly was approved by Dick Cheney who served as secretary of defense under Bush Sr.[33]

33. What's all this talk about the petrodollar?

In August 1971, after President Nixon decoupled the dollar from its gold backing, a short-lived dollar panic started. The US understood that a lack of trust in the dollar was going to be a problem. Clearly, some other backing for the dollar was urgently needed. Nixon and his secretary of state, Henry Kissinger, feared a decline in the global demand for the US dollar was imminent.

It was Kissinger who came up with the brilliant idea of asking Saudi Arabia to agree to only sell oil in dollars and to invest a portion of these dollars in US Treasuries.[34] The money that the US government received in this manner—known as petrodollars—would then be recycled back into the American economy. This arrangement required a constant increase in the supply of dollars.

After a series of confidential meetings, Saudi King Faisal in 1974 accepted the American proposal, but only if it remained a strict secret.[34] In return, Saudi Arabia was to receive any military protection needed for its royal family and its growing oil empire. The agreement now provided a double whammy as the large dollar surplus from oil sales was used for both buying US Treasuries and arms sales. The US also promised to help the country build a modern infrastructure (using American companies, of course). The US had found a way to protect its economic hegemony. The other OPEC countries followed suit, and by 1975 all of OPEC had agreed to sell their oil in dollars. They also agreed to purchase weapons in the billions of dollars with their newly gained fortunes.[12]

Not coincidentally, countries that opted to sell their oil for currencies other than the dollar have met serious opposition. In 2000, Iraq converted all its oil transactions under the Oil for Food program to euros.[35] When the US invaded Iraq again, three years later, oil sales from this country switched from the euro back to dollars. Iran created its own oil bourse in 2008. It started selling oil in gold, euros, dollars, and yen. Venezuela supported Iran's decision to sell oil for euros.[36] Libya presented a threat to the petrodollar in 2010; Muammar Gadhafi wanted to create a pan-African currency called the gold dinar that could be used for their oil transactions, and to replace the euro-pegged West and Central African Francs, guaranteed by the French treasury.

According to released e-mails from Hillary Clinton's server, the potential switch to a gold-backed currency was a primary motive for the NATO intervention in Libya in 2012. After the Libyan revolution, the country continued to sell oil in dollars.[37] Syria had switched to euros in 2006 to deal with US sanctions, and the US has been seeking a regime change ever since.[38]

The only real challenge for the petrodollar trade would be if the BRICS—Brazil, Russia, India, China, and South Africa—decided to drop the dollar for international trade, or if they purchased natural resources instead of US Treasuries with their dollar holdings, since this would endanger the dollar's world reserve currency status. This is exactly the exit strategy China and Russia seem to be aiming for right now. In recent years, the Russians have sold most of their US dollar holdings, while they quadrupled their gold position.[39]–[41] The Chinese have stopped buying extra US Treasuries since 2010 and sold quite a bit in 2016, while they have imported and invested in huge amounts of gold.[42] These developments signal the first stages of the US dollar's decline.

34. So oil is a weapon in a financial, economic war?

A recent example of financial economic warfare was the sudden crash of the oil price and value of the ruble after the annexation of the Crimea by Russia in March 2014. Soon after—in less than six months, from July 2014 to January 2015—the price of oil dropped from $103 to $47 per barrel. The speed of this drop could not be fully explained by supply and demand fundamentals, since the oversupply of oil was only 2 million barrels per day in a 90-million-dollar-per-day market.

Some market commentators said it reminded them of the Cold War era, when the US and the former Soviet Union competed not only militarily but also by trying to play the economy. Because the former Soviet Union became increasingly dependent on food imports in the 1980s (especially grain), the export of oil had to bring in enough dollars to buy food. The US decided to use its influence on Saudi Arabia, and US president Ronald Reagan allegedly persuaded King Fahd to expand the supply of oil, according to Reagan's son Michael.[43] Whether this agreement actually took place is still unknown, but King Fahd did meet Reagan both in 1984 and 1985, to conclude multibillion US dollar arms deals.[44] Saudi production grew from 4 to 9 million barrels per day between 1984 and 1991, resulting in an oil price plunge from $65 to $35 per barrel (in real value). This sharp rise in Saudi production came on the back of an earlier 1980–1983 production plunge that sought to compensate for an oversupply of oil as a consequence of the earlier oil crises.[45] The drop in oil prices would soon prove to be a fatal attack for the former Soviet Union, with oil income dropping from $300 to $130 billion, the economy shrinking by 20%, and the Soviet Union itself collapsing in 1991.[46] The fact that Saudi Arabia in 2014 again increased its oil production, albeit by 0.5 million barrels per day, has fueled rumors of a new economic war against Russia.

In economic warfare, the aim is to capture or otherwise control the supply of critical economic resources or destroy a country's currency. The United States understands better than anybody else that a country can be more severely damaged by destroying its economy than by bombing its infrastructure. The recent collapse of the oil price since the end of 2014 has led to the collapse of the Russian ruble. CEO Herman Gref of the Russian Sberbank, the country's biggest bank, confirmed that it had come under a financial economic attack in December 2014.[47] In an interview he disclosed that a foreign-based attempt was made to provoke a bank run during the ruble crisis. About $6 billion had been withdrawn from the Sberbank in a single day after a massive disinformation attack, with people receiving text messages saying Sberbank was facing problems paying out deposits. Thousands of SMS messages were sent, including a large number of mailings done from foreign websites.

In recent years we have seen more examples of such economic warfare. In May 2015, the US had a number of high-ranking FIFA officials arrested in Switzerland in connection with a bribery case.[48] The US action was designed to pressure FIFA, according to Alastair Crooke, a former UK MI6 official. He is one of the few individuals who have spoken openly about the purpose of this kind of financial and economic warfare.[49] In April 2015, several US senators wrote a letter to FIFA, 'urging it to consider removing Russia as host of the 2018 FIFA World Cup because of its role in the Ukraine crisis and occupation of Crimea.'[50]

But there is an even more serious weapon in the arsenal of the United States. In the book *Treasury's War*, the tool to exclude countries from the dollar-denominated global financial system is described as a 'neutron bomb.'[51] When a country must be isolated, a 'scarlet letter' is issued by the US Treasury that asserts that such-and-such bank is somehow suspected of being linked to a terrorist movement or involved in money laundering. The author of *Treasury's War*, Juan Zarate, a former senior US Treasury and White House official and the chief architect of modern financial warfare, writes that this scarlet letter constitutes a more potent bomb than any military weapon. With Ukraine we have a substantial geostrategic conflict taking place as part of a geofinancial war between the US and Russia with the European Union countries in the middle.

35. Does natural gas play a role in this new 'cold war' too?

Natural gas is increasingly used as a political weapon. In Europe this became clear after the 'gas crisis' of 2005–2006 and in 2008–2009 when Russia briefly turned down the gas tap to Ukraine because of a dispute over a steep upping of the gas price by Russia, and because Ukraine had not paid its natural gas bills of $1.3 billion and $4.2 billion on time.[52],[53]

EU leaders became concerned that gas received from Russia through Ukraine could be at risk, despite the Russian government stating that EU gas deliveries were not in danger. So a large policy shift emerged within the EU. Many felt the Russians were no longer a secure partner even though they had been one in the 40 years before, without missing a single delivery.

Shortly after the 2005 'gas crisis', the EU countries proposed two new southern gas pipeline routes from the Caspian Sea to diversify away from Russia. Both were intended to transfer gas from Azerbaijan to Europe.[64] This would start to endanger Russia's control over the EU, as well as bring large economic gains for Azerbaijan, Turkmenistan, and Uzbekistan.

Russia responded quickly and announced the new North and South Stream gas pipeline projects. They would form alternative routes to Europe for Russian gas, one via the Gulf of Finland into Germany, and the other via the Black Sea through Bulgaria. The North Stream raced ahead as German, French, and Dutch companies BASF, E.ON, Gasunie, and

---

64 Nabucco would go through Bulgaria onto Austria and Germany, and TAP enters Greece and onto Italy.

Engie were eager to collaborate in the lucrative deal, with construction of North Stream completed in 2012, despite initial Swedish opposition. Now Ukraine is increasingly being bypassed, and transit through Ukraine will largely end in the coming years as North Stream ramps up.[54]–[56]

Shortly thereafter, Russia proposed a second North Stream, and in 2015 Shell, BASF, E.ON, Engie, and Austrian oil and gas company OMV agreed to invest and develop the pipeline.[57] Several EU member states are trying to halt the project. Poland's regulator has raised monopoly objections, and nine EU member states, formerly part of the Soviet Union, have sent an official letter of objection to the European Commission. The US vice president Joe Biden, whose son was involved in a shale gas project in Ukraine, stated the pipeline was a 'bad deal.'[57] Yet the existing consortium, supported by the governments of France, Germany, and the Netherlands, pushed forward, and it looks like the project will go ahead.[58],[59]

Russia has been less successful with South Stream. Construction of the pipeline for Russian gas directly to Bulgaria through the Black Sea began in 2013, but soon ground to a halt as the EU parliament voted to end the project, and Bulgaria had to pull out, after the annexation of Crimea. Russia abandoned the project in response to EU sanctions, even though it had already spent $4.7 billion.[60] The EU seems to be gaining the upper hand now in the Southern part of Europe, with construction having commenced of its TAP pipeline project that connects Italy to Turkey via Greece. The amount of gas from Azerbaijan will at best replace 6% of Russian exports to the EU-28, if the pipeline is completed by 2020.

Russia has not given up on its plans, and it launched an alternative to the South Stream pipeline to Turkey in

2014, under the name Turkish Stream. The plan was put on the back burner after Turkey shot down a Russian fighter plane in Syria in 2015. But after an official apology from Turkey, the pipeline deal was revived.[61],[62] The next step, to connect Turkish Stream to the European Union countries, is underway, as Greece, Serbia, Hungary, and FY Republic of Macedonia are about to sign a memorandum for construction of the so-called Tesla Pipeline from Turkey to Austria. In essence 'Tesla' would deliver Russian gas, thus consolidating Russia's economic control over the EU as its key gas supplier.[63][65]

Today the EU still imports 35% of its gas from Russia via pipeline, and this dependence will only grow. Gas production peaked in the UK in 2000 and has declined by 65%, in Denmark it peaked in 2008 with similar declines, and in the Netherlands we saw a peak in gas production in 2013 with a 25% decline afterwards.[66] Only the gas output of Norway is stable and expected to remain steady up to 2030. All in all, production in the gas countries of the EU is estimated to drop by another 100 BCM or 40% by 2035 from 2015 levels.[64]

65   The exemption of European investment sanctions on Gazprom, the main Russian Gas company, despite the majority of oil & gas companies being targeted following the annexation of Crimea by Russia, underscores this economic dependence.[65]

66   The reduction in the Netherlands followed a Dutch court ruling which forced output of the Giant Slochteren gas field to be capped at half of its production, over concerns of an increasing number of small earthquakes. The smaller gas fields in the country are already nearly depleted and Slochteren provides 60% of total output.[66]

Chapter 4 – Peak Oil Revisited: The End of Cheap Oil

Our analysis suggests there are ample physical oil and liquid fuel resources for the foreseeable future. However, the rate at which new supplies can be developed and the break-even prices for those new supplies are changing.
– International Energy Agency, 2013[42]

All the easy oil and gas in the world has pretty much been found. Now comes the harder work in finding and producing oil from more challenging environments and work areas.
– William J. Cummings, ExxonMobil company spokesman, 2005[62]

It is pretty clear that there is not much chance of finding any significant quantity of new cheap oil. Any new or unconventional oil is going to be expensive.
– Lord Ron Oxburgh, a former chairman of Shell, 2008[63]

Introduction

Peak Oil is the point in time when the maximum production of oil in the world is reached. After that point is reached, oil supply can no longer grow, it remains stable for a few years at best, and typically starts to fall unrelentingly. It is often confused with oil depletion and running out of oil altogether. Shell geophysicist M. King Hubbert in the 1950s was the first to figure out why this phenomena of an oil field's production rising, peaking, and declining occurs the way it does. Thanks to his in-depth understanding of the physics of oil production and observations of past discoveries and production levels, he was able to predict future production trends.[67] Hubbert's first prediction, from 1956, that US easy-to-extract conventional oil production would peak around 1970, turned out to be accurate.[1] He was vindicated only long after the fact, as attacks by US government bodies and geologists on his prediction continued several years after the peak had occurred. Even the United States Geological Survey in 1974 still published reports stating that production would grow to 20 million barrels per day by 1985 after a few years of declining production (in reality in 1985 production has declined to 9 million barrels per day).[2] US oil production after the peak continued to decline with smaller peaks and declines until 2008, after which it started to grow again. The recent change has not been because of a reversal in the peak in conventional oil, but rather due to an explosive rise of unconventional shale oil using hydraulic fracturing and

---

67   The physics behind peak oil relates to oil being a liquid that sits under pressure in porous rocks underground. As oil is extracted pressure declines up to a point that its flows slow down and production peaks, it then further declines until oil no longer flows out of the rocks. Production techniques can go only so far to halt this process.

horizontal drilling techniques. US oil production thanks to shale oil increased by 74% from 2008 to 2015, towards 9.4 million barrels per day.[3] In contrast the highest peak for conventional oil still stands at the 1970 value of 10 million barrels. Hubbert's second prediction from 1956, that world oil production would peak by 2000, turned out to be too pessimistic.[1] While he has been largely correct for conventional oil production which seems to have peaked in 2004, he did not foresee the 1970–1980s oil crises that would lower oil demand, nor did he anticipate substantial growth in unconventional and costly-to-extract oil production.[2][68] Similarly to Hubbert, many others have been attacked because of their predictions that the peak in oil production would occur in the period of 2010 to 2020. Many, including the authors in our previous book in 2008, have so far been too pessimistic about the growth of unconventional oil, although our earlier prediction—'the peak in conventional oil will be reached within a few years'— is still holding up in retrospect.[69] As we discuss in this chapter, there is no reason to be complacent about future oil supplies. In 2015/16 US oil production rapidly declined again to around 8.5 million barrels per day, largely due to the excessively high cost of shale oil extraction.[3] In 2005, the United States Department of Energy published the Hirsch report about the need to prepare.[4] It stated, 'The peaking of world oil production presents the United States and the world with an

68 The definition of conventional oil used in the book includes medium and light oil in onshore and shallow offshore waters, and excludes deepwater as well as tar sands, shale oil, and heavy and extra-heavy, due to the substantial differences in effort needed to retrieve these latter 'unconventional' forms of oil.

69 In our book de Permanent Oliecrisis from 2008, we wrote: 'In the optimistic case oil production will grow at a low pace up to 2015 towards 88 million barrels per day. Yet it is more plausible that the peak in conventional oil will be reached within a few years, because of which total oil production may begin to decline.'

unprecedented risk management problem. As peaking is approached, liquid fuel prices and price volatility will increase dramatically, and, without timely mitigation, the economic, social, and political costs will be unprecedented. Viable mitigation options exist on both the supply and demand sides, but to have substantial impact, they must be initiated more than a decade in advance of peaking.' To date the US government's 'International Energy Outlook', published annually by the Energy Information Administration, does not expect a world oil production peak before 2040, as it assumes oil can be supplied increasingly from expensive unconventional oil sources and production methods. Their low oil price scenario requires, for investments to flow, an oil price of $76 per barrel and relies heavily on OPEC, while their high case necessitates a whopping $252 per barrel if OPEC fails to perform and as even far costlier sources than today need to be extracted.[5] Whether this is feasible in light of investment needs and the already fragile world economy remains to be seen. The cost of oil has a profound effect on economic growth, as high oil prices can lead to economic stagnation or even contraction.

36.   Why hasn't world oil production peaked yet?

We need to find new oil fields to continue producing enough oil, yet the volume of discovered oil has declined every decade since the 1960s. Real warning signs for peak oil appeared in the 1980s, as global annual production levels rose above the number of discovered barrels. For several decades we have been eating into our stock of oil in the ground. Yet, to the surprise of many, total oil production has continued to increase substantially in the last 10 years,[70] despite the growing gap between discoveries and production.

The answer to this mystery relates to what type of oil we are talking about. Oil production in recent years increased only due to rapid growth in unconventional oil production from shale oil, tar sands, extra-heavy oil, and deep-sea oil fields. The total sum of conventional easy-to-produce oil has declined since at least 2005, from a peak of around 68 to a level of 60 million b/d (figure 8).[6]–[9] This is oil that is extracted in either onshore or shallow offshore reservoirs, up to 300 meters depth, and without special techniques such as hydraulic fracturing or heat injection. So conventional easy-to-extract Peak Oil has arrived already, just as we predicted in 2008.

Other sources of oil cost up to $90 per barrel and require far greater infrastructure and labor investments. The increasing need for oil companies to invest in unconventional oil to expand production is a sign of oil companies' belief in the end of easy cheap-to-extract oil. Ex-Shell CEO van der Veer has explained the end-of-easy-oil concept very well

70   Global and country level oil production figures are publicly available from the International Energy Agency (IEA) of the OECD and the U.S. government's Energy Information Administration (EIA).

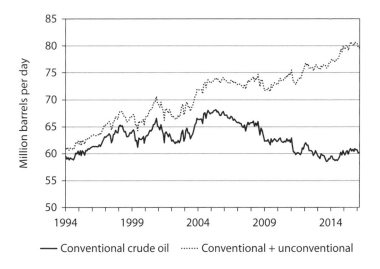

Fig. 8. Production of conventional and unconventional crude oil from January 1994 to late 2015. The difference is caused by substantial growth in the production of deep-water oil, tar sands, shale oil, and to a lesser extent extra-heavy oil. Sources:[6]–[9]

in several publications. This statement was published in a Council of Foreign Relations publication in 2008:

> So what is the easy oil? We don't know exactly but we wouldn't be surprised if this oil would peak somewhere in the next 10 years. Then there's the unconventional oil. We've got the unconventional oil onstream. But what the total is, we are not that precise.[10]

In addition to conventional and unconventional oil, three other sources of liquid fuels, not shown in figure 8, exist; natural gas liquids (NGL), biofuels, and refinery gains. All have added to total world liquid fuel supplies in the last two decades. NGLs are produced as a part of natural gas extraction and are primarily converted into petrochemicals and to a lesser extent gasoline. Total production of NGLs grew from 6.4 to 10.5 million b/d between 2000 and 2015. Biofuels have a non-fossil origin and are derived from corn ethanol in the United States, sugarcane in Brazil, and sugar beets and rapeseed in the EU, and grew from 0.3 to 2.6 million b/d in the same period. And refinery gains are increases in volume when processing of crude oil because the resulting products weigh less per unit of volume then crude oil doe. In a simplified sense the crude oil molecules when split are 'spread' across more litres. Refinery gains grew from 1.5 to 2.4 million b/d between early 2000 and the end of 2015.[6]

In 2006, the International Energy Agency (IEA) predicted that by 2030, demand and supply could reach 116 million barrels per day, as an extrapolation from previous years.[11] Global demand for crude oil grew an average of 1.8% per year from 1994 to 2006. After reaching a high of 86 million barrels per day in 2007, however, world oil consumption decreased in both 2008 and 2009 due to the international financial crisis. At present oil demand is projected by the IEA to increase by 20% from 2007 levels, to reach 104 million barrels per day by 2030, a 0.8% average annual growth, mainly due to demand growth in the transportation sector.[12] Over 65% of the oil used in the United States, and 59% worldwide, is used in passenger vehicles powered by internal combustion engines.[13],[14] This Revolution is therefore of particular importance if we are to mitigate the effects of peak oil.

37.   When will we reach world peak oil?

Over 70,000 oil fields have been discovered worldwide. Most have been brought into production, and many have already passed their peak production level and produce less every year. Since the finiteness of oil expresses itself in a pattern of growth, peak, and decline, the oil industry has to add new oil fields continuously to current world production, otherwise total production will fall. Where possible, oil companies try to stave off the peak in already producing fields, or even temporarily reverse the decline by using new production techniques, but this can only do so much. Thus the higher production is, the more new fields need to be brought into production every year just to maintain production levels.

The International Energy Agency (IEA) in its latest 2015 outlook estimated that by 2030, 45% of 2014 production will need to be replaced by oil from new fields. This is because conventional oil production from large existing fields, once they have peaked, declines on average by 6.5% per year, and for smaller fields the decline is above 10% annually.[15]–[17] It's why the oil industry invests globally over $600 billion every year to maintain oil production.[18][71,72] In 2005 the global oil industry invested only $130 billion in exploration and development, and about $100 billion on average during the 1990s.[19],[20][73]

71   Conventional plus unconventional oil production in 2015 based on IEA data excluding biofuels, natural gas liquids, and processing gains.

72   Capital investment in 2014 was close to 700 billion US dollars and dropped to about 580 billion in 2015.

73   Nearly a 1000 drilling rigs were in operation internationally in 2014 outside of China, Russia and the U.S. versus 600 in the year 2000, based on Baker-Hughes rig count data. In North-America rig counts were in the low 200 versus over 1500 in 2014, prior to the oil price decline into 2015.

Conventional oil production is declining in a growing number of oil-producing countries. The onshore production in 25 out of 39 countries is declining where production either still is or once was above 100,000 barrels per day. And similarly, offshore oil production in 24 out of 34 countries with either current or past significant historic production levels is in decline (see figure 9).[8][74] Recently onshore China has been added to this group with production first stabilizing and now declining as the supergiant Daqing has reached its declining phase.[21][75]

The last large new cheap-to-extract province was the North Sea. Exploration started there in the 1970s due to oil security concerns and increasing oil prices. The large discoveries of 60 billion barrels were quickly exploited and were already exhausted by the late 1990s, when production peaked at 6 million barrels per day in the UK, Norway, and Denmark. Today these countries only produce 2.9 million barrels per day together.[22]

Despite recent high oil prices, no new North Sea oil has been discovered with oil that can be extracted at a cost less than $20. Total oil discoveries have continued to decline despite oil price growth, and in the last five years we produced four times more oil than we discovered, with about 43% of discovered oil coming from deep-sea costly-to-produce fields.[23] (See question 23).

We can also see that high oil prices have not done much to squeeze more oil out of existing fields. Both onshore and offshore production in countries that have peaked each declined by over 2 million barrels per day between 2008

74 The offshore values exclude deepwater oil to reflect conventional oil production

75 Offshore oil field production in China is still growing and reached 1+ million barrels per day in 2016.

Fig. 9. Production in million barrels per day stacked for peaked onshore (a) and growing onshore (b) oil region by country, and for peaked offshore (c) and growing offshore (b) region by country. Offshore values exclude deep-water, onshore values exclude extra-heavy oil, oil sands, and shale oil.[6],[8],[9],[24]

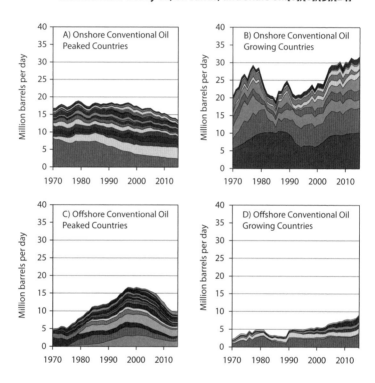

and 2015. Only a temporary respite is visible for 2014/15 in Norway and the UK due to recent one-off discoveries that came online (figure 9c). And the continued conventional oil production growth in other countries was not sufficient to compensate for declining countries, as shown in the earlier figure 6. Conventional production has peaked and is declining.

Fig. 10. World oil production from deep-sea oil fields (300+ meters), tar sands, shale oil, and extra-heavy plus heavy oil Sources:[8],[9],[24]–[26]

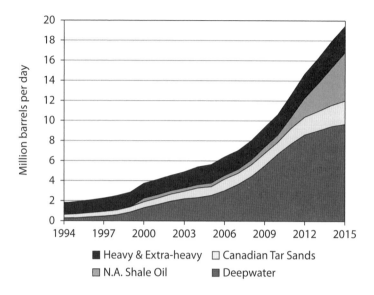

So the only way to continue postponing peak oil is to continue production growth from unconventional sources, and to do it faster than declines in conventional oil fields. Today the world produces about 19 million barrels per day from costly-to-extract unconventional fields, and this would need to grow by at least another 2.5 million barrels every year, so as to grow oil production (figure 10). A key source of oil in recent years has been shale oil produced in the United States and Canada that comes from shale rock reservoirs that provide much less easy flowing oil than conventional oil fields.

The amount of oil in the ground from unconventional sources is vast, especially tar sands, shale oil, and extra-heavy oil that doesn't flow, as summarized in table 5 on page 157. Since unconventional oil comes at a high price, the drop in oil prices between 2014 and 2016 has moved the target for peak oil much closer.

In reaction, almost 70 oil projects, especially deep-sea and tar sands, have been postponed, which would otherwise deliver an additional 2.8 million barrels per day by 2022.[27],[28] The total investment in oil and gas exploration and development dropped by about $250 billion in 2015/16. The last time this happened was after the 1980s oil crisis.[29],[30]

In Canada, the largest tar sand producer, capital spending dropped by 62% or $31 billion from 2014 to 2016.[31] Balance sheets of oil majors in 2014 clearly show the need for at least $80 per barrel to break even, and require $50 to $60 to stay cash flow neutral.[18]

The energy analysis firm Rystad Energy—based on their database of 65,000+ oil fields—has calculated that oil prices have to be over $60 per barrel to grow oil production again.[30] At best, oil production will remain stable in the next years. Since it takes seven years from investments in deep-water, tar sands, or extra-heavy oil to first new field production, the investment downturn has a long-lasting impact.

There are two possible scenarios that could unfold for world oil production now:

– The optimistic scenario heralds that high oil prices will lead to renewed high investment in unconventional oil. Since a majority of new projects require $60 to $80 per barrel to warrant investments and provide profits,[32] oil prices will have to remain consistently above $60 per barrel for this scenario to happen.

– The pessimistic scenario is that production will decline as investments continue to fall short. The declines in conventional production will become too large, with especially the Middle East disappointing in growing its production.

Table 5. Estimated global oil resources and reserves from USGS data (in billion barrels)[33]–[39][76]

Resource Category	Already Produced	Discovered Oil left in Place	Proven Reserves
Conventional crude excl. Middle-East*	941	-	414
Middle-East Conventional Crude*	377	-	810
Conventional oil total	1,318	-	1,224
			Technically recoverable
Deep-water oil	27	130	36
Tar Sands	10	2,501	243
Shale oil	8	6,867	341
Extra-heavy oil	3	1,938	258
Unconventional oil total	48	11,436	878
NGL	n/a	n/a	n/a
Yet to discover NGL estimate	n/a	372	139
Potential resource-to-reserve EOR**	n/a	1,205	598
Yet to be discovered crude oil	n/a	1,208	478
Conventional undiscovered and EOR	n/a	2,413	1,076

* Middle East includes Saudi Arabia, Iraq, Iran, Kuwait, UAE, Syria, Yemen, Oman, Qatar, among others.

** Enhanced Oil Recovery (EOR) based additions from discovered oil left in place to technically recoverable

76   The resources and estimated reserves of oil in the ground we use are globally assessed by the United States Geological Survey (USGS) of the U.S. government. Statistics are also available from the German geological Institute BGR, in the BP Statistical Review of World Energy, and the World Energy Council in their survey of World Energy Resources.

Based on these trends in global oil production, we could see large volatility in oil prices till at least 2020. But within five years, the Tesla Revolution could be well underway, with many electric and hybrid cars flooding car markets. What many do not realise is that this is a self-accelerating process. Once electric cars start to have a major impact on oil demand, companies will make fewer investments in oil supply. Less oil will come to the market, which will spell a new era where the oil market is no longer driven by rising oil demand but the reverse.

38.   So the era of cheap oil is already over?

Perhaps Chevron's CEO Dave O'Reilly captured it best when he warned that the 'era of easy cheap oil was over' as early as 2005, in an interview with *Fortune* magazine.[40] Increasingly, oil producers need to drill into several kilometer-deep waters, reach difficult fields in distant locations, and tap into unconventional sources of oil which require a substantially greater effort to transform into usable form. We have entered an era where hundreds of shale oil wells are necessary to obtain the same production as a single oil field with just a few wells.

The key issue is not so much the actual quantity of oil underground but the ability to obtain sufficient oil at an affordable cost. We can imagine an oil field underground to be a sponge-like structure of tiny rocks interspaced with oil that sits under pressure. The process of drilling an oil well is nothing more than tapping a straw into the reservoir, thereby making the oil flow into it and to the surface. Existing pressure in an oil field initially provides for natural flow.[77] Deeper drilling thus requires a far bigger steel pipe,

---

77   The art of oil drilling: At increasing depths a tunnel as big as a basketball in width is drilled with a diamond or tungsten carbide drill at the end of a long chain of steel pipes, which form the base of the well. Drilling takes place by constantly circulating so-called drilling mud – consisting of water, clay, and chemicals, to cool down the drill and add pressure onto the drill-bit. Once the desired depth is reached the steel is encased in place with cement on the outside by pumping cement into the entire steel column, so as to prevent contact with underground water, Typically this is done in several stages, each with a slightly thinner steel pipe that fits into the previous one. Once the segments are in place another small pipe is lowered alongside the entire column wherein the oil flows, together with a set of explosives as part of this production pipe, which when set off at the bottom provide for holes through which the oil can flow. In case of hydraulic fracturing additional water is pumped down into the pipe to add pressure, and more heavier explosives are used to increase the sideway fractures into the rock.

and in the sea also a far thicker steel pipe since pressure builds with greater sea depths. Since the straw is longer, it also takes more drilling time, which all add up to the far greater cost of deep-sea drilling than close to the surface oil fields. The average capital cost of new deep-sea fields (at 300 to 1,500 meter depths), and ultra-deep sea fields (at 1,500 meters or greater depths), has been estimated to be $1 and $3 billion in 2016, respectively.[41] Today half of all new discoveries and about a quarter of oil production growth comes from deep-sea and ultra deep-sea oil fields.[7],[32]

Since oil flow is caused by pressure, the amount that flows out of the ground is not constant. When a well is first drilled, oil flow increases as more and more oil is sucked through the porous rocks, production grows, and reaches a maximum. The well then produces at its maximum level for a period of several weeks to years, after which production declines because pressure drops too much. The well has peaked and production continues to decline, up to the point that it stops because the flow becomes so small that it is no longer economical to continue production. Since an oil field consists of several to many wells, it follows the same pattern of peak production. The only difference is that placing more or fewer wells (straws) will affect the speed at which oil is sucked out, and thus the height of the peak and the onset of production decline.

This pattern of growth, maximum peak production, and decline can only be stalled, and at times temporarily reversed, by inducing additional pressure on the oil in the well or by breaking the rocks through which oil flows. Techniques include water-flooding which entails injecting water in another point near the existing well to increase pressure, as well as steam and carbon dioxide injections, also referred to as enhanced oil recovery (EOR), or hydraulic fracturing by pumping water into the well under pressure to break the rocks.[42] The application of these techniques is estimated to require $60 to $80 per barrel of oil for most regions.[32]

At the end of an oil field's life, even after including the pressure increasing techniques, only a portion of the oil is recovered from between the porous rocks. It varies from 5% to 90% for a given oil field, with so far on average 35% globally.[43] The value created from oil fields has grown during the history of the oil industry, as technology and economically favorable situations have led to an increase in this so-called resource recovery rate.[43] More and more oil resources are turned into extractable reserves, yet recently, growing the recovery rate comes at a steeper price than before, and the added benefits are diminishing.

So the key drivers of growth in global oil production are unconventional oil sources. These include:

– **Tar sands**, that is oil that seeped to the surface and became mixed with sand, is either mined or processed with high heat to 'boil' out the oil, or extracted underground using steam injection. An oil price of $50 to $70 per barrel is needed for steam injection projects. Mining-based projects require an even higher oil price of $80 to $90.[44]

– Very thick oil, also called **extra-heavy**, needs either steam injection to thin it, or is pumped out and then diluted with lighter oil from other fields. It is mainly found in the Orinoco belt in Venezuela as well the Middle East. This source needs $70+ per barrel for investments to flow.[45]

– **Shale oil**, which needs hundreds of wells to get to sizeable production levels, is developed with hydraulic fracturing and horizontal drilling. In the US shale oil costs now range between $40 and $90 per barrel, depending on the field and its geology, with a majority requiring $60+ oil prices.[46]

— **Deep-water oil** fields, which are oil fields in the sea at either 300 to 1,500 meters of depth (deep-water) or below 1,500 meters (ultra deep-water). Due to the substantially greater pressure at these depths, substantially more steel piping is required, and production is far costlier to develop. These fields require $60+ per barrel in Brazil and over $80 per barrel in other countries for investments to happen.[47]

The most notable development, because of its breathtaking speed, has been the shale oil revolution in the United States. From 2008 to 2014 it added over 5 million barrels of oil daily to the production mix, bringing total US oil production back up to 10 million barrels per day.

Because of the increasing need to grow total oil production (from deep-sea oil fields, via enhanced oil recovery, and from unconventional sources), the cost to bring any additional barrel to the market has grown substantially, from about $15 in 2000 to $70 in 2015. Cost projections based on the type of new oil available reach up to $130 per barrel by 2030, assuming that the investment will be available and oil companies are able to turn available resources into extractable reserves so as to continue production growth.[5],[12]

39.   How did shale oil grow so quickly?

The main source of unconventional oil in recent years has been shale oil resource plays in the United States. The large ramp-up to a production level of 4.5 million barrels per day in 2015 came as a complete surprise. In themselves shale oil and gas were not unknown resources; in 1821 a shallow shale gas well was brought into production.[48] Geologists have understood since the nineteenth century that many such plays existed, as shales formed the initial starting point of oil and gas formation. And even politicians were aware of the potential with president Richard Nixon in a 1973 presidential address calling for a program to develop commercial extraction of shale oil in the US.[49] Yet its rapid development surprised even insiders.[50]

Shales are the so-called 'source rocks' in which oil was formed from organic matter that was deposited during periods of warming 50 to 150 million years ago. A large part of the oil formed in these rocks migrated through porous cracks and microholes into 'reservoir rocks' of conventional oil fields, which formed a more concentrated pool enclosed by a cap rock that trapped the oil in place. In the underground layers where such cap rocks were not in place, the oil continued to migrate slowly to the surface, where bacteria either ate it over millions of years, or it became mixed with sand, forming what we now today call the tar sands of Canada.

What geologists also knew was that shale oil was difficult to extract because of the low porosity of these rocks. Many vertical wells would have to be drilled with little payoff, were it not for two key innovations made in the last 15 years: the perfection of horizontal drilling, and the utilization of hydraulic fracturing via explosives and chemicals to open the rocks underground for oil to flow. These innovations made it possible to extract a lot of oil from shale in a short time, at a cost of about $70 per barrel around the year 2010.

Shale deposits are situated deeper underground than conventional oil fields, and in formations that were more horizontally stretched out and not as thick. The start of shale oil and gas exploitation at scale in 2005/06 became possible because of new techniques like hydraulic fracturing or 'fracking', which entails pumping a dense liquid down an oil or gas well at pressure to crack open the rock.[78] Hydraulic fracturing had already been utilized for over half a century; in its first application in 1948, a mixture of leftover Second World War napalm mixed with sand was pumped down the well to pressure the rocks open to allow gas and oil to escape.

Testing hydraulic fracturing in the 1980s in shales was first done by a US company called Mitchell Energy. The company was after shale gas and first drilled many uneconomic wells in the Barnett Shale, but it managed to rapidly reduce the cost of fracturing as it drilled over 589 wells between 1982 and 2000.

Only a few years later, large-scale exploitation started after prices jumped to over $50 around 2005. This increase in oil prices made it feasible to translate the shale gas drilling experience with horizontal wells directly to shale oil.[79]

78 In conventional gas and oil extraction only one particular part of the casing is normally perforated using explosives with a lower charge, since no fractures need to be generated.

79 The first horizontally drilled well was completed already in 1929 but at great costs. Horizontal drilling up to 1000 meters became commercially viable in special cases in the US since the late 1980s at 25%-300% higher costs than vertical wells. In 1990 6% of total drilling expenditure in the US was spent on horizontal wells.[99] Yet it had not been commercially applied to shale plays.

Only since 2003 have horizontal wells been successfully and at large scale applied in the Barnett shale.

Today, three US regions provide over 85% of shale oil production. These are estimated to contain 33 billion barrels of oil in resource totals. Another 15 plays have been found to possess significantly large amounts; however, oil companies have not been successful in ramping up their production due to their different geological formations.[51]

In a situation of relatively low oil prices of around $40 per barrel, the majority of shale oil-producing wells return a loss. As a consequence, the drilling of new wells has virtually come to a halt in 2016, and oil production dropped by nearly one million barrels per day. Shale oil will thus only continue to provide significant quantities if oil prices rebalance to levels above $60.[52] Some producers are able to operate at lower prices due to cost cutting, but this is only the case for the best 'sweet spots.'[53]

For most companies, the majority of debt needs to be paid off in 2017 and 2018. If oil prices stay low, millions of oil workers will be laid off, and the US economy could face a severe economic hit.

40.   Why are experts and the media always wrong on oil prices?

Since oil prices are key to getting a picture of future oil supplies, accurate oil price forecasts are imperative, but for some reason experts always seem to get it wrong. In the late 1990s, oil prices dropped to $10 per barrel because of relative low demand and high supply. Institutes and analysts came out with predictions that low oil prices levels were here to stay. The *Economist* ran its now infamous 1999 edition cover 'Drowning in Oil' stating that the world was awash with oil and the price would range between $5 and $10 per barrel for the foreseeable future.[54]

This media report was the perfect contraindicator, with oil prices tripling in the same year as a consequence of OPEC and non-OPEC countries' production cuts, combined with higher than anticipated demand for oil. Similarly, forecasters saw continued increases in oil prices in the ramp-up to an all-time high in 2008. The Texas oil multibillionaire T. Boone Pickens in August 2008 stated that 'in two or three years, we're going to be at $200 a barrel – could be $300 a barrel—for oil.'[55] Only a few weeks later US oil firm veteran analyst Charles Maxwell came out with a similar $300 per barrel prediction.[56] They were also dead wrong. Oil prices dropped nearly $100 to $45 within a matter of months, in the wake of the global financial crisis.

Price forecasts seem to be more an art than a science. Extrapolations of the present situation into the future usually lead to error. To gauge the future, examination of the dynamic drivers that shape supply, demand, and prices is a must. Unfortunately the future of supply as well as demand holds too much uncertainty to make any accurate oil price forecast beyond a few fundamental tenets. First,

Fig. 11. Oil price development from 1861 to 2015 (in 2015 US dollars).[22]

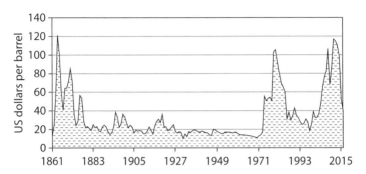

the end of cheap oil means that oil production cannot be expanded as easily as in the past. After the 1970s oil crises, there was a permanent decrease in oil demand growth in combination with large expansions of easy oil in many non-OPEC countries, which resulted in an oil glut.[57] Despite OPEC's attempts to raise prices by cutting production, they remained at or below $20 a barrel. This meant that there always was downward pressure on prices.

Now the environment has fundamentally changed. Since the year 2000, more and more producers have seen declines in cheap and easy-to-scale-up oil fields. The expansion of supply has become less flexible, as it necessitates larger investments and longer time horizons of seven years on average.[47][80] Second, expansion has become more costly due to which the minimum price is much higher than before. The minimum price level is influenced by the operational costs of already producing fields in the short

80   The only exception to flexibility is US shale oil, which can be increased rapidly within a six month to years' time-frame as adding a large number of wells can be done with a few months.

run of months. To maintain production in the long run the price needs to be high enough to provide revenue so that investments can be made to bring new fields into production in the medium run of years, without which production will drop substantially.[81] Both factors have resulted in a market environment with substantial upward pressure on prices where demand cannot be easily met. It takes the third fundamental tenet—that the oil price has a price ceiling because of demand destruction at high price levels—for prices under upward pressure to stabilize and decline, such as during the 2008 price hike and the ensuing financial crisis.

In that year, oil usage in road transport remained stable and even saw a drop of 0.7% in 2009, similar to the 1979/80 oil crisis.[58] More importantly, these high prices also fundamentally change the usage of oil, as witnessed following the oil crises, due to changes in behavior and refocused attention on substitution by companies. Since the ramp-up in prices to 2008, oil for heating is being phased out at an accelerated pace in many countries, China has been building a coal-based petrochemical industry, biofuels have grown in importance, and oil use in road transportation declined every year from 2008 to 2013 in the EU countries due to efficiency and demand destruction.[59],[60]

---

81   The analyst firm Wood Mackenzie has estimated in 2016 for existing production were investments have already been made that – from an operation cost perspective – a 40 USD per barrel oil implies 2 million barrels per day are uneconomic to produce, at 30 USD per barrel up to 5.3 million barrels per day are uneconomic, and at 20 USD per barrel close to 12.8 million barrels per day of production are uneconomic.[100]

41.   What are the oil market implications of rising oil prices?

The high oil prices in 2005–2014 provided some of the largest transfers of wealth between oil-consuming and oil-producing countries in the history of industrialized countries. Within just this time span of 10 years, a total of $8.5 trillion, equivalent to about 1.2% of global economic output, was transferred via oil sales to the 14 OPEC oil-producing countries in the Middle East, Africa, and South America.[61][82] High oil prices provide a large economic stimulus to oil-exporting economies.

At the same time, oil-importing countries suffered, as they had to pay the price of this transfer of wealth, mainly in smaller economies whose oil imports at the price levels from 2008 to 2014 resulted in a large burden on the import-export balance. The 'PIIGS' countries of Portugal, Ireland, Italy, Greece, and Spain, saw their oil import budgets balloon from 2005 to 2008 by 75% to a joint $120 billion, which only reversed with the 2014/15 oil price drop.[62]

A key effect of the growth of petrodollar exchanges between countries has been the expansion of the number of dollars used in international trade. Because of the petrodollar system, the United States has benefitted substantially as the price of the dollar remained stable despite the country's domestic financial problems and trade deficit. As explained earlier, no other country has the financial luxury of running imbalances at such high levels without heavily impacting the purchasing power of its citizens.

---

82   Wealth transfer values are net of oil production costs as provided by OPEC in its annual statistical bulletin.

Beyond changes in oil wealth distribution, high oil prices also provide an impetus to invest and develop alternative energy sources. A considerable number of alternatives to oil are close to or at commercial market levels at oil prices above $100 dollars. Since 2005, largely due to the impact of oil prices, research budgets in energy have quadrupled across the world.

The extraction of large amounts of additional unconventional oil in general, including shale oil, tar sands, extra-heavy, and deep-water, requires an oil price above $60 per barrel.[63] Beyond analyst calculations we can see this from the only success story for shale oil outside of the US so far, in Argentina, where production started in 2013 and has reached 20,000 barrels per day due to a strategic effort from the government, which includes a minimum $67 guaranteed oil price level.[64],[65]

The other main countries where shale oil production could unfold in the near-term future are Mexico and Russia. The first successfully producing shale oil well was drilled in Russia in 2016 in the Bazhenov formation in Siberia,[83] which potentially holds as much recoverable shale oil resources as the entire United States.[66]–[68] In Mexico, shale oil blocks will be auctioned off to foreign oil and gas companies in 2017 for the first time since its oil industry nationalization, which, given promising geology, will kick-start shale extraction.[69],[70] China also has large shale resources but these are extremely difficult and thus too costly to extract.

---

83   The well was drilled to 2300 meters depth and produced 330 barrels of oil per day, which is in the same order of magnitude as US shale oil plays.

42.   Is future oil production at risk?

The largest uncertainties for oil stem from the secrecy around oil reserve estimates in the OPEC countries. This issue arose after the nationalization of oil companies in Saudi Arabia, Kuwait, Iran, Iraq, Venezuela, and the United Arab Emirates.

Since the 1970s, these countries operate free of direct Western influence, whereas in the old days Western oil companies were in control. For example, British Petroleum—the private oil giant from the United Kingdom—used to be called Anglo-Persian and operated as a sole player in Iran. Nowadays, only national insiders are aware of the true state of their reserves since the companies are not stock-listed and issues of oil reserves are state secrets. They are shrouded in suspicion as a consequence of unwarranted historic increases.

In 1984 Kuwait published 50% of additional reserves on its books. Since production quotas for OPEC countries at the time were based on the reserves of a country, adjusting its books upward enabled a country to increase its production.[71] Not long after Kuwait magically inflated its reserves, Venezuela did the same, in 1985, and two years later the United Arab Emirates, Iran, and Iraq followed, each with higher adjustments than the other. Kuwait increased the reserves on its books from 67 to 93 billion barrels, after which Iran reported that it possessed 94 billion barrels. The United Arab Emirates couldn't stay behind, and it reported a massive tripling from 33 to 97 billion barrels, and the rulers in Iraq thought 100 billion barrels would be representative of their oil wealth. Saudi Arabia, possibly the only country which actually carried out a real reserve evaluation, upgraded its books in 1988 from 170 to 260 billion barrels.[61]

In total the OPEC reserves from all countries combined suddenly grew by 306 billion barrels within five years. Today OPEC reserve numbers are taken at face value, even though there is no way to verify their accuracy. The only clear acknowledgement within OPEC that the books were indeed cooked has come from the Saudi Sadad-al Husseini who, during the Oil & Money Conference of 2007 in London, presented his view that today OPEC oil reserves should indeed be lowered by 25% or 300 billion barrels.[72] Since Sadad-al Husseini was responsible for exploration and production at Saudi Aramco, the state oil company of Saudi Arabia, before his retirement in 2004, he probably understands these matters better than anybody else.

The second challenge is the uncertainty over political and financial stability in many OPEC countries. Large amounts of financial capital and know-how are needed to expand oil infrastructure and unlock new oil fields, either from within a country, or from foreign oil companies and financiers with deep pockets. The exploitation of a medium-sized 100,000+ barrels per day onshore oil field in a remote location or a shallow offshore oil field in water up to 300 meters depth can easily require a sum of $250 million.[41],[73],[74]

The OPEC countries typically operate from a national oil company that provides the government with 80% or more of its state budgets. This constrains available investment and sometimes reduces it to zero when national oil companies are used as cash cows to fill holes in the budget. In all OPEC countries except Saudi Arabia, Qatar, and the UAE, outside funding is needed to supply the tens of billions of US dollars to bring new large oil fields into production. Countries thus need to set financially favorable rules that assure that any investment will deliver a return on investment. And they also need to convince investors that they are stable enough for at least a single decade, before money starts flowing. If the risk is too high, companies are probably unwilling to invest, or they depart after a period of learning their lesson the hard way. Such issues have arisen in recent years in Venezuela, Nigeria, and Libya, that led to multibillion dollar losses in asset write-downs for oil companies including Shell, ExxonMobil, and Total.[75]–[78]

43.   How much can 'cheap' OPEC production grow?

We need to replace about 3% of world oil production annually just to keep production steady.[12] It has been estimated that this decline rate of the world's oil production has grown substantially in the last decade and will continue to be high for the foreseeable future. Fatih Birol, the head of the International Energy Agency:

> Even if demand remained steady, the world would have to find the equivalent of four Saudi Arabias to maintain production, and six Saudi Arabias if it is to keep up with the expected increase in demand between now and 2030.[79]

The majority of this new oil, according to the same IEA, needs to come from OPEC countries. The world needs about 20 million additional barrels per day by 2030, to compensate for declines. So OPEC oil production needs to grow from 32 to 39 million barrels per day.[12][84,85]

The biggest upside in recent years can be found in Iraq and Iran. Production in Iran is growing for the first time since 2004, thanks to the lifting of international sanctions in 2015 by the United States and the EU over its nuclear energy program. Now oil export from Iran is no longer restricted to mainly China, and international investments can begin to flow into the country.[80] Production still has room to almost double to reach Iran's highest production of 6 million

---

84 Growth within OPEC mainly has to come from Iran, Iraq, Venezuela, and Libya, and to a lesser extent Saudi-Arabia, Kuwait, the United Arab Emirates, and Qatar and Nigeria.
85 The values exclude Natural Gas Liquids, as these are produced alongside natural gas production.

barrels per day reached in the 1970s, prior to the 1979 Iranian revolution.[6]

The development in Iraqi oil production in recent years is nothing short of remarkable. In its entire pre-'democratic' history, Iraq only managed to produce 3.5 million barrels per day, while 2015 production reached a good 4 million barrels per day despite the rise of ISIS and constant bombing by US and European nations over many years.[6]

Rising production comes mainly from the oil fields in the Kurdish region in northern Iraq (KRI). After the removal of Saddam Hussein by the United States in 2005, the Kurdish authorities—without official approval from Baghdad—opened all its territories to foreign oil companies. In the last 10 years in the KRI, about 15 oil fields have been discovered with 7 billion barrels in reserves.[81] As of 2015 it produced 500,000 million barrels per day, from zero in 2008, and growth to 1.4 million barrels per day is expected by 2020.[81]

The development of oil fields in Iraq's central and southern oil-rich regions remains plagued with difficulties, mainly because a $40 to $50 per barrel oil price leaves too little government money to pay back foreign oil companies under current contracting terms.[12][86] Thus, even though Iraqi oil production will undoubtedly continue to grow, it is unlikely that the Baghdad government's plan to reach 7 million barrels per day by 2021 will pan out.[82]

---

86   The contracts the Iraqi government in Baghdad offers are so called production service contracts, these oblige the government to reimburse a foreign companies capital expenditure and operating costs, and also pay a small variable fee for each produced barrel. Since development is capital intensive, and nearly all of the government's budget comes from oil revenues (also used to pay the foreign oil firms), a 40 to 50 US dollar oil prices means that too little oil money is left for sustained investments.

Quite uncertain is the oil future for both Libya and Venezuela. Libya caught the world's attention because of the downfall of its nefarious dictator, Muammar Gaddafi, brought down in 2011 by several armed domestic groups supported by weapons and airstrikes from the US and European countries. Ever since, oil production has collapsed from about 1.5 million to less than half a million barrels per day, as domestic groups are fighting over control of the country.[6] Since 2013, the situation has worsened as the terrorist group ISIS has established a stronghold in the country's oil-rich region of Sirte, from which it is only slowly being expelled.[83]–[85]

Venezuela holds vast deposits of unconventional extra-heavy oil, of which 220 billion barrels are potentially recoverable.[87] Production started in 1998 from four projects jointly producing 450,000 barrels per day, developed between PdVSA—the state oil company—and Chevron, BP, Total, and Statoil.[88] A second wave of six projects was implemented after the oil industry was nationalized by Hugo Chavez in 2007, with dozens of oil majors participating including Repsol, CNPC, Chevron, PetroVietnam, Eni, Rosneft, and Lukoil. Extra-heavy production in 2015 stood at 1 to 1.3 mb/d varying by estimate.[26],[86] Since 2004 total oil production has ranged between 2.8 and 3.2 mb/d, and new extra-heavy projects thus only compensated for production declines in mature conventional oil fields.

---

87 The late President Hugo Chavez in 2005-06 initiated the Magna Reserva project to examine the extractability of these resources. A large seismic assessment and 134 appraisal wells were drilled on the basis of which 220 billion barrels were found to be of sufficient quality to be technically extractable.[101]

88 Extra-heavy oil in Venezuela is produced by pumping the oil out with heavy-pressure equipment, and subsequently diluting it by blending in lighter oils.

While PdVSA seeks to increase extra-heavy oil production by another 3 mb/d, this is not going to happen anytime soon. Such a large increase necessitates investments of over $100 billion, but PdVSA has no capital available, and the country is close to default. Because of massive money printing after the strong decline of oil prices in 2015, the Bolivar currency is losing its value, with inflation reaching 145% in 2015, and the IMF anticipates hyperinflation.[87] While PdVSa is highly profitable, it provides only about 35%–40% of the state's revenues. It has resorted to high debt loads to finance its expansion. In 2006 the company was $3 billion in debt, which has grown to $25 billion in 2010 and $46 billion in 2014.[88],[89]

44. How important is Saudi Arabia in this respect?

The biggest wild card is perhaps Saudi Arabia, the largest producer of cheap and easy-to-produce oil, with production of over 10 million barrels per day in 2015. The Saudi's state that their oil production can rise to 12.5 million barrels per day, and be maintained until at least 2050.[90],[91]

It is striking that the country still produces 10 million barrels per day, similar to the early 1980s, over 30 years ago. A key reason for the limited rise in Saudi production is their high decline rate in existing oil fields. In 2003 a Saudi Aramco official said wells in Saudi's existing fields were declining at a rate of 5% to 12% per year.[92] Drilling of additional wells in existing reservoirs has stabilized decline rates to 2%.[93][89]

The biggest uncertainty for Saudi Arabia is the future of Ghawar, the largest oil field in the world both in size and production. Ghawar is the driving force behind Saudi wealth and a determining factor for future oil production in the country and the world. Persistent rumors are that Ghawar may be close to the end of its life, yet it could equally continue to produce high rates for the foreseeable future. This famous field was discovered in 1951 and rapidly delivered over half of all oil produced in Saudi Arabia. The last somewhat reliable data show that the field in 2007 produced 5.1 million barrels per day, nearly 6% of total world oil production.[94]

Ghawar, due to its size – its surface is about as big as Trinidad and Tobago or five times Luxembourg – can be divided

---

89 Saudi-Arabia needs to drill many more wells each year now to keep their production steady, and has close to 80 oil rigs today in operation versus just 20 in 2004.

into different regions, which each can be seen as separate oil fields from a production perspective.[90] As of today, over 5,000 wells have been drilled in the entire Ghawar oil field.[95]

The expectation from within Saudi Arabia has been less optimistic than the expectations of official energy agencies in Europe and the US. In 2007 the International Energy Agency (IEA) and US Energy Information Administration (EIA) estimated that Saudi Arabia could produce at least 16 million barrels per day by 2030.[96],[97]

The former head of exploration and production of Saudi Aramco, Sadad-Al Husseini, has called such high production expectations impossible. In an interview after the Oil & Money Conference in October 2007, he stated: 'Many international organizations have assumed that Saudi Arabia can double its production towards 20 million barrels per day; that is an unrealistic expectation.'[98] It took many years for these organizations to substantially adjust their expectations. In 2015 the IEA estimated that 13 million barrels per day will be produced by Saudi Arabia in 2030, a sign reality is starting to set in.[12]

Sadad-al Husseini also indicated that expectations for OPEC countries to grow oil production substantially above 30 million barrels per day are unrealistic as well. The IEA in contrast still anticipates that by 2030 the OPEC countries will produce 46 million barrels per day.[12] In 2015 OPEC produced exactly 30 million b/d, excluding Angola and Ecuador, the two countries that joined OPEC in 2007 after Sadad-al Husseini's statements.[91]

90   Ghawar spans across 230 km by 30 km in surface area.
91   Excluding natural gas liquids.

Chapter 5 – Climate Change and the World of Energy

It's amazing how many people don't actually believe this is a serious threat, even though it is unequivocally the most serious threat that humanity faces across the board, in every country of the world. We are essentially playing Russian roulette with the ocean and atmosphere.
– Elon Musk, 2015

The real wild card in the pack is increasing vulnerability from resource scarcity and climate change, with the potential for major social and economic disruption. Unless we take action on climate change, future generations will be roasted, toasted, fried, and grilled.
– Christine Lagarde, 2013

There are two doors. Behind door #1 is a completely sealed room, with a regular, gasoline-fueled car. Behind door #2 is an identical, completely sealed room, with an electric car. Both engines are running full blast. I want you to pick a door to open, and enter the room and shut the door behind you. You have to stay in the room you choose for one hour. You cannot turn off the engine. You do not get a gas mask. I'm guessing you chose door #2, with the electric car, right? Door #1 is a fatal choice—who would ever want to breathe those fumes? This is the choice the world is making right now.
– Arnold Schwarzenegger, 2015

Introduction

Melting ice caps, at least in the Northern Hemisphere, rising sea levels, and record temperatures are hot topics in our daily news.[1] Even large storms and droughts are more and more often linked to climate change. Although many doubt manmade changes could turn the tide of rising global CO_2 and temperature levels, climate accords are now leading to guidelines for many governments, especially in the G20 countries. When we add the growing smog problem, especially in the Asia, it is easy to understand why so many countries are now supporting investments in alternative forms of energy like wind and solar. It is even leading to a revival of the nuclear industry. In this chapter we try to present climate facts as objectively as possible. To report, not to campaign. It shows the world of energy is highly connected to the world of climate change, so anybody wishing to understand the future of energy should at least study the latest developments in climate science and CO_2 reduction measures.

45.   Is the climate really changing?

Many of the climate news reports come from data collected by a dozen institutes and research organizations. Global temperature is updated monthly by the American NOAA from satellite and sensor data.[92] In their monthly 'state of the climate', NOAA described the situation for July 2016 as:

> For the 15th consecutive month, the global land and ocean temperature departure from average was the highest since global temperature records began in 1880 [...] The July 2016 combined average temperature over global land and ocean surfaces was 0.87°C (1.57°F) above the 20th century average, besting the previous July record set in 2015 by 0.06°C (0.11°F).

These values are averages for the entire planet relative to an estimated twentieth-century-average temperature of 15.8°C (60.4°F). A large part of the world's land and oceans, especially in the Northern hemisphere, are thus experiencing several degrees warmer temperature, while others are actually colder. We can see that this change is part of an increasing trend. It was substantially colder in the 1900s and temperature has steadily increased since (figure 12).[93]

Datasets from satellite, sensor, or ground measurements for other aspects of climate show other trends:
– Data on rainfall is also tracked for different parts of the globe. Globally, precipitation since 1900 has fluctuated between minus 90 to plus 60 millimeters per year, and there are no clear upward or downward trends. Deserts

---

92   US National Oceanic and Atmospheric Administration, See (www.ncdc.noaa. gov/sotc/).

93   In the 1960's newspapers reported about the risks for global cooling.

Fig. 12. **Global average land-ocean temperature changes from direct measurements from 1880 to 2015, relative to the mean average from 1951 to 1980. Source: NASA GISTEMP[2]**

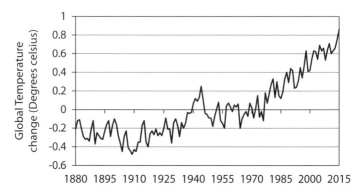

typically get only 25 to 200 mm per year, whereas rain-forests receive intense 1,500 to 3,000 mm of rainfall. If we zoom in on specific regions we get mixed patterns: Australia is getting wetter and southern Europe is getting drier.[3]

- Data on sea-level rise is available globally and per coastal region, since the height of the sea and its change differs across the globe. The sea level at New York is one meter higher than that in the Bermuda area.[94] The average global sea level increased by 29 centimeters from 1870 to 2016, of which 9 centimeters between 1993 and 2016.[4]

- The Arctic and Antarctic regions are covered by ice caps, and measurements are available of their size in millions of square kilometers. Since 1965 the arctic ice cap started to melt and dropped from 13 to about 10.5

94 Sea level changes are different per coastal region due to ocean currents and physical forces such as gravity. See for a map of differences per coast: tidesandcur-rents.noaa.gov/sltrends/sltrends.html.

million km² on average during the year.[95] Antarctic sea-ice increased from about 11.7 to 12.6 million km² between 1979 and 2015, while Antarctic land-ice is quite stable.[5]

- Greenland is also covered by an ice sheet that has lost an average of 3 meters of its 2.5 km-thick land-ice layer since 1900.[6] Since the year 2000, the loss is accelerating, and a total 3 billion metric tons of ice, or 0.12% of the total ice mass, has melted. About 0.01% of the ice is now lost every year.[7],[8]

- The number of extreme weather events have tripled since 1980, from 200 to over 600 major loss events,[96] resulting in substantially more weather damage.[9],[10] In the Atlantic basin, records show a doubling of hurricanes formed per year.[11] Large heatwaves and increased incidents of flooding are also more frequent, like the Russian summer of 2010, which exceeded records of hundreds of years and caused 30% in grain harvest losses, and the highest rainfall in Germany ever measured flooded Prague and Dresden in 2002.[9],[11]

The climate thus seems to be changing. Especially the last 15 years look different than the climate in the 1980s to 1990s. But the main question is, what is causing this change?

95   The sea-ice in the arctic and Antarctic fluctuates substantially. In case of the arctic between summer in June to October and winter in November to February. In 2012 this fluctuation was between about 3.5 and 15 square kilometers.
96   Major loss events are defined by re-insurance giant Munich Re as 'events have caused at least one fatality and/or produced normalized losses ≥ US$ 100k, 300k, 1m, or 3m (depending on the assigned World Bank income group of the affected country)'.

46.   What is the role of carbon dioxide (CO2)?

The starting point of our climate is our sun. Without its rays, our planet would be a frozen ball of ice. It takes about eight minutes for sunlight to reach the earth. A small portion is reflected away by earth's atmosphere and clouds, but the largest part reaches the surface. As the surface heats, it also emits back sunlight into the atmosphere as infrared radiation.[97]

Now the role of gases in the atmosphere comes into play, including carbon dioxide, water vapor, and methane, which can absorb the infrared radiation from sunlight. Similar to warming the skin on your arm, these gas molecules heat up in the atmosphere, yet they also quickly re-emit the infrared radiation in a random direction, bouncing into another gas molecule. Because of this process of absorption and re-emission, the sun's heat is 'trapped' longer in the atmosphere. Without greenhouse gases, it would be 33 degrees Celsius colder on earth. The more such gases in the atmosphere, of which carbon dioxide is the most important[98], the more heat remains 'trapped' and the hotter the earth.

Greenhouse gases are not the only absorbers of the sun's heat. The water in the oceans warms up as the sun's rays

97   Sunlight enters the earth initially as visible light and ultraviolet rays, which heats up the ground and part of that energy is re-emitted into the atmosphere as infrared radiation. That is why greenhouse gasses mainly come into play when the sunlight is re-emitted, and not initially when sunlight enters the atmosphere.
98   There is about 100,000 times the carbon dioxide in the atmosphere then methane, about 70% of the extra carbon dioxide stays in the atmosphere up to 300 years, about 20% for 1000 years, and about 10% for up to 400,000 years, versus in total about 12 years for methane.[36],[37] Whilst there is 5x more water vapour than carbon dioxide in the atmosphere, it only stays there for a few days and thus does not accumulate from emissions.

penetrate meters deep. Yes, water can hold lots of heat. The air temperature difference above water and land also explains why there are strong winds on beaches into land, as wind is the movement of air from colder to hotter places.

The oceans also distribute heat and can exchange carbon dioxide with the atmosphere like plants can.[99] Ocean surface water is heated around the equator in the Pacific Ocean and Indian Ocean, and currents bring the warmer water along Africa up to northern Europe, which results in mild winters in northern Europe. The United Kingdom lies as far north as Quebec, Canada, yet the temperature rarely drops below the freezing point. Seawater, after its arrival in the North Atlantic, cools down and sinks, and flows back at deeper ocean depth to the Indian Ocean and Pacific Ocean.

The climate on earth is thereby formed by interactions between the sun, air over land and oceans, the ocean waters, and greenhouse gases like carbon dioxide.[100] Daily weather is formed by variation within the climate due to differences in temperature and humidity across land and sea, driven by day-to-day sunshine variation.

99 Carbon dioxide dissolves in sea water, and chemically turns into carbonic acid (similar to baking soda), which turns into bicarbonate and hydrogen ions, which stays in the deeper ocean layers for a long time. It also is taken up by small micro-organisms (phytoplankton) which when it dies sinks to the bottom of the ocean.

100 These factors are influenced by several others such as the rotation of the earth and earth's orbit, the currents in the oceans, volcanoes, and plate tectonics in the very long-term.

Fig. 13. Temperature and carbon dioxide variation up to 11,500 years ago based on Roman calendar (top), up to 800,0000 years ago from present (middle), and up to 65 million years ago (bottom).[12]–[24]

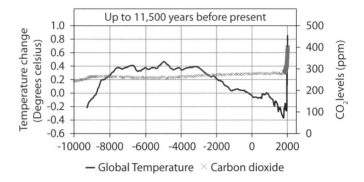

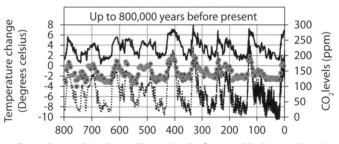

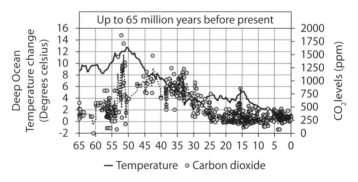

47. Hasn't climate always changed?

The world's climate has always been changing, yet the speed of change and its direction have not always been the same. In the last 10,000 years, the climate has been remarkably stable, with average global temperature estimated to have ranged between -0.4 and + 0.4 degrees Celsius, relative to the average between 1961 and 1990.[24][101] Carbon dioxide levels remained quite stable during this period, at around 280 parts per million, up to the recent industrial age, when they both shot up in unison (figure 13).

We also have good records for temperature and carbon dioxide, up to 800,000 years old, from the gaseous composition of air trapped in ice cores (figure 13). Every three meters deeper you will find air that is about 300 years older. With ice cores to three kilometers deep, we can go far back in time.

Ice cores show that the remarkably stable warm period of the last 10,000 years is actually an exception. About 16,000 years, ago we were in a cold 'glacial' period, up to 10 degrees Celsius colder than today, with polar ice extending across North Europe and North America.[102] Such glacial periods are historically the norm, as for about 80% of time in the last 800,000 years we experienced glacial periods with temperatures at least 4 and often 8 degrees Celsius colder. Only every 100,000 years or so did warmer periods of

101   We can make reasonably accurate historic records based on various indirect measurements including tree rings, fossils, ice cores, pollen, corals, shells, soil and sediments. see: www.ncdc.noaa.gov/data-access/paleoclimatology-data/datasets.
102   Glacial periods are not the same as ice ages which are much longer periods of millions of years in time defined by the existence of large ice sheets on northern and southern hemispheres. Today we are in a warm interglacial period within an ice age.

about 10,000 to 15,000 years takes place, at about the same temperature as in the twentieth century.

Ice core data provides the best evidence of the link between carbon dioxide and temperature, as they follow each other closely in this period, with a delay between initial warming and mass carbon dioxide release established at 200 years.[25] This interaction and the cause of warm periods to glacial ages can be explained in six steps:[26]

1. Warming starts because of changes in how the earth rotates on itself and spins around the sun (the orbital cycle). Every 110,000 years or so it is just in the right position for the northern part of the globe to get a lot more sunlight (northern Europe, Russia, Greenland, Arctic, Canada, northern United States).

2. The warming melts large parts of the ice sheets that go up to Europe, Canada, and North-China, and releases warm fresh water (no salt), which changes the ocean currents that make northern Europe warm, and results in a warming of the Southern Hemisphere and oceans.

3. Now a temperature feedback turns the oceans into a net emitter of carbon dioxide. Deeper ocean water flows upwards as the ocean heats up, which contains a lot of dissolved carbon dioxide and dead microorganisms composed of carbon. The carbon-rich water gets closer to the surface, a lot of fish grow which feed on the dead microorganisms that breathe out carbon dioxide like humans, and this happens at such mass rates that the ocean starts to release carbon dioxide instead of absorbing it.[27][103]

103 Also the bicarbonate increasingly turns back into carbon dioxide as the heat-chemical balance changes. In a way this is like baking bread, where baking soda when exposed to an acidic ingredient (like brown sugar), turns into carbon dioxide which forms bubbles in bread.

4. The large amounts of released carbon dioxide result in more heat being trapped which accelerates warming and turns ice ages into warmer periods. Carbon dioxide levels shoot up from 180 parts to 280 parts per million in the atmosphere in these transitions.[28] Now the earth enters a warmer period like we have experienced in the last 10,000 years.

5. As the earth's orbit continues to change during the 10,000 to 15,000 or so warmer years, the Northern Hemisphere receives less sunlight and becomes gradually colder. Snow is more frequent which turns into ice, and as snow and ice reflect more sunlight than bare earth, it accelerates the process, leading to a colder climate.

6. A glacial period starts when large amounts of added carbon dioxide from the earlier warmer release get absorbed again because of cooler ocean waters (dropping levels back to 180 parts per million). The glacial period only ends when the earth's orbit is again just in the right position to let the sun melt the majority of the vast sheets of ice that have formed in the north (step 1 above).

The cycles of glacial and interglacial warm periods were regular in the last one million years, yet we can go further back—much further—thanks to crystals and sea floor 'soil.' Sediments composed of decaying shells and sea life, as well as geochemical compositions of crystals, allow for the reconstruction of temperature and carbon dioxide records up to 550 million years ago.

Deep time records are much fuzzier and only provide general trends at geological timescales. We can see that about 2.5 million years ago, the world started to experience an ice age, as the polar caps became permanently covered with ice, which kicked off the glacial – interglacial dynamics that we are still in today. The earth before the ice age was much warmer. It cooled, very slowly from a temperature of 12 degrees Celsius warmer than today, nearly 49 million years ago, about 17 million years after the dinosaurs became extinct (figure 13). It was so warm at that time that palm trees grew in polar regions.[29] Proxy records of carbon dioxide up to 65 million years ago show a less distinct relation to temperature, partially still unexplained. Carbon dioxide levels rose slowly from around 400 to 1,600 parts per million from 50 to 65 million years ago, and began a long descent that stopped at about 200 parts per million reached around 10 million years ago.

A part of the explanation is the net uptake of carbon dioxide by plants in the ocean and land to produce oxygen. As plants decay, the carbon dioxide contained as carbon in their roots, stems, and leaves gets stored in soils and ocean sediments, and also results in added formation of oil, gas, and coal. By slowly taking more and more carbon dioxide out of the atmosphere, the amount of 'trapped' heat declined, and the climate cooled down.[30]–[33]

48.   What is different this time?

The rate of carbon dioxide in the atmosphere has been increasing rapidly for over 100 years. The key difference today is that this is not due to ocean and climate dynamics but is due to the burning of carbon by humans stored in oil, gas, and coal, as well as deforestation and cement production. The earth's atmosphere everywhere now contains carbon dioxide levels above 400+ parts per million (ppm).[34] The last time the atmosphere contained this level of carbon dioxide was about 13 million years ago. Based on some of the most solid ice core records available, carbon dioxide levels in the last 800,000 years fluctuated between 180 and 280 ppm up to around the year 1750.[28]

In one way the rise in carbon dioxide is positive, as we will no longer have to worry for a long time about cold glacial periods with temperatures eight degrees lower. From modelling studies of the sun's light intensity, carbon dioxide levels, and glacial periods, we know that such a large temperature drop, even if humans do not add any further extra carbon dioxide into the atmosphere, will likely not happen for another 100,000 years.[35] There simply is too much heat added in the atmosphere trapped in carbon dioxide molecules, and when the gas is rapidly injected in the atmosphere, it is calculated that about 20% of it will stay for 10,000 years and 10% for up to 400,000 years, because of how the gas exchange with plants, oceans, and sea life works.[36],[37]

Yet this upside is not really worth the risk now because we enter uncharted climate territory. The best, or should we say worst, analogy to today's situation—of a rapid increase in carbon dioxide in a very brief period—is the so called 'Paleocene-Eocene Thermal Maximum' or PETM which happened 56 million years ago. A rapid release into the atmosphere of 1.1 billion metric tons of carbon per year occurred over several thousand years at that time, resulting in a temperature increase of 6+ degrees Celsius over 10,000 years, after which it took about 150,000 years to get back to the original temperature.[38]–[42] Today's injection of carbon dioxide is happening about 10 times faster than during the PETM. The climate conditions were also vastly different, as temperatures were 10 degrees higher than today, and the atmosphere contained over 1,000 parts per million of CO_2.

So we do not have a clear idea what will happen due to the rapid increase of CO_2 emissions from fossil fuels into the atmosphere. Thousands of researchers are trying to understand from past records and the earth's physics how the future could pan out, often creating computer simulations of climate interactions.

The sum of their work is published in a summary report every five years by the Intergovernmental Panel on Climate Change (IPCC).[43] In figure 14 we have reproduced a few of the forecasts from this large body of experts from the 2013 IPCC report, the most recent one.[104]

The conclusions in the report come from computer models that still lack many key feedback loops that can accelerate or decelerate warming or cooling, since our understanding in these areas is limited. We still for instance

104 The IPCC conclusions are based on simulations with four scenarios for carbon dioxide concentrations in the atmosphere up to 2100: stabilising at 450 or 550 parts per million, and growing to 750 or 1250 parts per million. In news coverage on climate change, typically, the worst scenario with 1250 parts per million (called RCP 8.5), assuming rapidly increasing fossil fuels across the 21st century, is used to describe the effects of changing climate, on issues such as extreme weather, sea-level rise, and so on.

Fig. 14. IPCC scenario outcomes for temperature on earth up to 2100 for temperature relative to 1986–2005 levels (B), sea ice extent (C) and surface permafrost (D) both in the Northern Hemisphere. Scenarios are from different models by varying carbon dioxide levels (A), including RCP 8.5 (circles), RCP 6 (diamonds), RCP 4.5 (crosses), and RCP 2.6 (triangles), plotted with historic data (black line).[43]

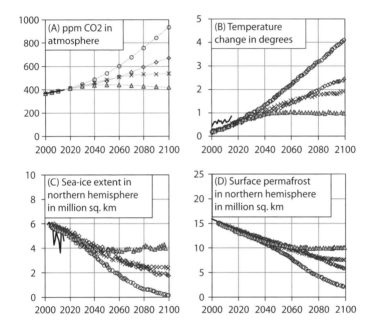

don't fully understand how clouds are formed,[105] or at what temperatures the permafrost will melt in Russia and Canada (how much carbon dioxide and methane trapped within it will be released into the atmosphere, accelerating

---

105 Higher thinner clouds let light through and instead trap some of the infrared radiation emitted by the earth and radiate it back down, thus warming the surface of the earth.

temperature increases). Because of these uncertainties and with no existing precedent of such a rapid accumulation of carbon dioxide in all history, Elon Musk has called emitting massive amounts of carbon from fossil fuels 'the dumbest experiment in history.'[44]

49.   What are the risks of our 'climate experiment'?

Since we do not have a good enough grasp of the process of climate change, there is a major challenge in understanding the risks of it. So far, most projections are based on the recent 2013 IPCC outlook that looks at developments till 2100, which for instance indicates sea levels will likely rise between 0.4 and 0.6 meters.[45] Yet this does not rule out that there is a small probability, of a few percent, that sea levels will rise several meters by 2100, which would pose a major threat to tens of millions of people.[106]

Of course impacts do not stop by the year 2100 but will continue for centuries to millennia, as carbon dioxide stays in the atmosphere for a very long time.[107] The risks for the twenty-first century, described below, are the conservative version of what may happen as opposed to the long-drawn-out threatening or sometimes called *scaremongering*, but highly uncertain, scenarios.

The sector which is most serious, and ahead of anybody else, in assessing risks of extreme weather events is the reinsurance industry: the companies that insure the insurance companies. These multibillion-dollar reinsurance giants have the daily task of examining losses caused by floods, storms, droughts, and heat waves. The German company Munich Re estimates that about 850,000 people lost their lives as a consequence of extreme weather from 1980 to 2014.[46] It also reported that total losses due to weather

106 The most extreme sea-level rise ever recorded within 100 years was 5 meters.[94] According to a probability analysis of upper limit sea level projections a rise of 1.8 meters or higher has a probability below 5%.[95]

107 About 70% of the extra carbon dioxide emitted stays in the atmosphere up to 300 years, about 20% for 1000 years, and about 10% for up to 400,000 years.[36],[37]

events in the early 1980s were below $50 billion per year, yet today they frequently exceed $100 billion (inflation adjusted).[10],[47] Although global loss forecasts are not yet available, German insurers have determined that extreme floods in Germany normally occur every 50 years, and that this might change to every 20 years, resulting in an overall doubling of flood damage losses.[48]

Extreme heat scenarios are a great concern especially for the Middle East and North Africa (MENA), as they add to an already hot climate. Temperatures in MENA countries are anticipated to increase by 2 degrees Celsius by 2050, with the number of extremely hot days and hot nights increasing from 16 to about 40 per year by 2050.[108] These are days with temperatures of 40 degrees Celsius and sometimes surpassing 50 degrees, such as in the summer of 2016 in the Middle East.[49] Temperatures during hot nights are expected to rise from below to above 30 degrees Celsius.[50],[51]

Big impacts are anticipated for places that will sink below sea level when sea level will rise by 0.5 meter by 2100. An assessment of the 180,000 islands on earth shows that many thousands could be lost, in the absence of human intervention or natural sand extensions.[52],[53][109] These include famous island groups like Kiribati, the Maldives, Seychelles, Solomon Islands, Polynesia, and Micronesia. On larger continents, about 22% of coastal wetlands could be lost.[54]

Studies of flooding impacts for almost 140 large low-lying port cities show 40 million people and $3 trillion in infrastructure (5% of the global economy) could be

108 Values based on IPCC RCP 4.5 scenario (550 PPM CO- peak in 2040) relative to average of 1986 to 2005.
109 Cyclones bring in rubble and earth that can extend island areas.

impacted.[55][110] The combined impacts of extreme weather, sea level rise, and flooding may potentially lead to the need to relocate millions of people within or between countries.[51],[56]

The biggest impacts on food production will likely happen in the world's oceans, due to the combined effects of warming seawater and ocean acidification. Carbon dioxide dissolves in seawater and chemically turns into bicarbonate (similar to baking soda) and hydrogen ions, which increases ocean acidity. The process removes a portion of the free calcium in seawater used by marine life to build shells and coral structures. Today the oceans are about 30% more acid than 100 years ago, and by 2100 could up to be 80% more acid if fossil fuel emissions continue unabated.[43],[57][111]

By that time it will become difficult for many species to form and maintain shells and calcium-based body parts.[58],[59][112] Early estimates show that ocean acidification and bleaching will lead to 10%+ reduced ocean life by end of the century.[60] Such evaluations are speculative, however, since it is unknown if ocean life can adapt rapidly enough to increasing acidification;[113] for instance,

110 See coast.noaa.gov/slr/ for an interactive sea level rise map of the United States.

111 Acidity is measured on a logarithmic pH scale. Every 0.1 drop in pH because of this is approximately a 26% increase in acidity. It is anticipated that the pH will drop further from 8.1 to 7.8 if emissions continue to grow and stabilise around 750 PPM Carbon Dioxide in the atmosphere (RCP 6.0) in the 22nd century.

112 Impacts are expected on coral reefs, starfish and sea cucumbers (echinoderms), the group of crabs and lobsters (crustaceans), and oysters and scallops (molluscs). Studies vary on the impacts of phytoplankton, the base of ocean life, with some estimating a positive and others a negative impact.

113 Little data and few studies have been undertaken of real life changes in pH on ocean life exist.

warm-water coral reefs may disappear entirely yet be replaced by sea grass and non-calcifying algae.[59],[61]–[63]

Increased warming of seawater especially impacts coral reefs via a phenomenon called coral bleaching. This is the effect of heat stress on algae normally present in coral structures which leave the coral that then turns white and dies off. The year 2016 was the worst coral bleaching year on record as 22% of the Australian Great Barrier Reef died off.[64] In the last few decades, bleaching events typically took place every 10 to 15 years, but by 2040 it is expected that they will become an annual occurrence.[65]

The jury is still out on food production on land. Crop yields grow due to heightened carbon dioxide (plants 'exhale' oxygen and 'inhale' carbon dioxide). Studies of real-field experiments show that grain productivity increases by 20% as carbon dioxide levels increase to 550 from 380 parts per million.[66] However, this effect is counteracted by temperature increases, which causes plants to go into water conservation mode and close tiny invisible holes in their leaves used for 'breathing'. The few real-field experiments of heightened carbon dioxide plus temperature show that the combined effects have a negative productivity effect. Since the biology of plants is not yet sufficiently understood, modeled global impacts vary widely, ranging from 40% declines to 40% increases in production for key crops like maize, rice, wheat, and soy, under high emission scenarios.[67]–[69]

50. What is the consensus among scientists?

Since climate change became a widely known cause for concern, several polls have been carried out among scientists. When scientists were asked to agree with the statement, 'Current changes in the climate have been caused primarily by humankind', 90% of all climate scientists agreed, while from the scientists not working directly on climate change but with a geological or weather-based knowledge base such as economic geologists and meteorologists, only about 45% to 60% agreed.[70],[71]

Not too much can be read into these polls because it is not clear what this consensus means for the future. The biggest disagreements amongst climate scientists and geoscientists relate to the strength of solar activity versus that of carbon dioxide on the earth's climate.

Solar activity has been measured directly from satellites since the 1970s, and prior to that from counting sunspots, dark spots that form on the sun when it is very active. Solar radiation strength varies in cycles of 11 years by 0.1%, so-called solar cycles, and much more over long time spans of thousands of years, due to the distance and angle of the sun to the earth.[72]–[74] Since only limited change in sunlight has been observed since the 1960s, it is difficult, if not impossible, to explain recent warming without the influence of carbon dioxide (figure 15).[75]

The degree of influence of carbon dioxide is even more uncertain than solar activity. A metric called *climate sensitivity* is used to describe how much warming over hundreds of years will occur if carbon dioxide levels in the atmosphere double. Since it represents the effects of various feedbacks, the climate sensitivity value is highly uncertain and its strength a matter of dispute. Climate sensitivity estimates

Fig. 15.   Solar radiation cycles and mean global temperature since 1880.[2],[76]

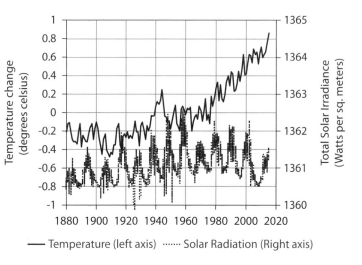

from the 15 most used climate models point to a rise in temperature of between 2.1 and 4.7 degrees Celsius, of which a 0.8 to 2.4 degrees Celsius increase is estimated within 70 years' time.[77]–[81] There is no clear consensus on whether temperature and climate impacts will be moderate or massive in the next 70 years, let alone by the year 2300.

51. Are climate change policies leading to lower coal use?

Burning coal releases the biggest amount of carbon dioxide into the atmosphere of all the fossil fuels. Getting rid of coal in the energy mix is thus often high on the agenda when countries set climate change policies, yet success in reducing coal use has so far been limited.

The first global agreement to reduce emissions was agreed by high-income countries in 1999, called the Kyoto Protocol. In this internationally binding agreement, 37 countries committed to reduce their carbon emissions by 2012 to below the level of 1990, including the EU countries, the United States, Japan, and Russia.

They succeeded but mostly as a consequence of the financial crisis of 2008 and low economic growth, and less due to government policies. Coal use in the 37 countries has been reduced since the agreement was signed in 1999 by 9% by 2012, and 17% by 2015, yet solely because of a drop in US coal use, coalesced by market forces and state level policies. Natural gas is increasingly substituted for coal in the United States. In the other 36 countries coal was maintained at its earlier levels.

At the same time, global emissions of carbon dioxide increased 52% from 1990 to 2012 due to growth in developing countries, not part of the agreement.

The Kyoto Protocol, while regionally successful, thus has not had a global impact on emissions. It did, however, set the scene for many countries to develop a regulatory framework to reduce carbon emissions.

An exceptional target has been set by the UK, a country which based on the Climate Change Act of 2006 instituted an 80% emission reductions target by 2050 in combination with a legally binding 2% annual reduction. The UK is phasing out all coal power stations and replacing them with natural gas and biomass pellet fuel imported from the United States.

China also set its own policies and targets in the country's 11th, 12th, and 13th Five-Year plans covering the periods 2005-10, 2010-15, and 2015-20. Its framework seeks to reduce the energy and carbon intensity of the economy by energy efficiency, heavy promotion of clean energy sources, and regional emissions trading.[82],[83] So far this has led the country to stabilize its coal usage at current rates, and there are even signs that coal use is on the decline.[84]

And under the Obama administration in the United States, the Clean Power Plan was proposed to Congress in 2015 to reduce carbon dioxide emissions by 32% from power plants relative to 2005 levels.[85] The plan would bind US states to this target and utilize carbon intensity standards to force older power plants out of action—substituting newer coal fuelled plants with natural gas power plants—and increase solar and wind energy. It has been halted, however, due to a US Supreme Court ruling.

The plethora of policies instituted by countries on their own just got bundled into a new international climate agreement, agreed at the end of 2015 in Paris. The agreement has the 'idealized' aim of keeping global temperatures below 2 degrees above pre-industrial levels.[114,115]

---

114 Typically the year 1900 is taken as a reference point for pre-industrial levels.
115 Countries that commit will need to set individual targets every five years called national climate action plans (INDC's). Before the conference 162 countries announced their first INDC's. Surprisingly, even Saudi Arabia contributed an INDC that targets energy efficiency, renewable energy, carbon capture and storage in its petrochemical industry, and the use of natural gas, with a target reduction of 130 million tonnes of CO_2 by 2030 (25% of Saudi current carbon emissions).[96]

The limited successes of the past, combined with the option to not ratify the agreement and the mixed ambitions of many countries doesn't point to a very promising picture for the Paris Agreement. As of October 2016, the agreement has been ratified since more than 80 countries have joined, accounting for over 55% of global emissions, including China, India, the US, European Union 28, Canada, and Brazil.[116] Nevertheless, individual country policies can drastically change the energy picture, as we have seen with the US and China recently. An unprecedented 2% global reduction in coal consumption occurred from 2014 to 2015 thanks to these two countries. We could well see the start of a new trend on the back of China's ambitious clean energy policies.[117] That's why the Chinese government is very active in stimulating this Tesla Revolution by giving large subsidies to Chinese citizens who buy EVs.

116 For the latest update see cait.wri.org/source/ratification/.
117 China invested more in clean energy than any other country in 2015.

52.   Will fossil fuels become obsolete?

The need to stay within a 2 degrees warming scenario, as agreed at the Paris conference, means that not all fossil fuels can be used and burned.[118] Researchers at London's University College have tried to figure out how much of current reserves need to stay in the ground to reduce the risks of climate change. Their rough estimate is that a maximum of 12% of current coal reserves, 48% of natural gas, and 65% of oil reserves can be extracted under this relatively arbitrary 2 degrees scenario.[86],[87] Similar results were discovered by the oil and gas analyst firm Rystad Energy.[88] To make this happen, gas production could only expand slightly towards 2050, coal extraction would need to be halved by 2030, and oil production needs to be reduced by 2% per year, and thus halved by 2050.

Canadian tar sands and extra-heavy oil production in Venezuela would be the first to fall victim, as well as the majority of new deep-sea oil discoveries, nor can unconventional shale and tight oil and gas expand substantially. The Carbon Tracker Initiative has estimated the financial impact of 'stranded assets' if governments would truly limit oil production expansion.[119] They think, between 2015 and 2025, 20% of future projects would need to be scrapped by oil majors like ExxonMobil, Shell, Total, Chevron, BP, Statoil, and Eni.[89] In total, $2 trillion in new coal mine and oil and gas field investments would need to be scrapped in the next 10 years to meet the Paris Agreement climate goal.[90]

118 Unless the emitted Carbon can be Captured and Stored (CCS), but at present this technology route is far too costly and little investment is taking place in CCS.
119 www.carbontracker.org/.

So this could become a serious financial threat for fossil fuel companies. The United Nations has set up an organization called the Portfolio Decarbonization Coalition (PDC) which is a growing coalition of institutional investors that seek to eliminate fossil fuel investments from their portfolios. Their 26 members jointly hold $3.2 trillion, an amount as large as almost 5% of the world's economy.[91] These include the Dutch pension fund ABP, and the Allianz Group, which will no longer invest in companies that derive over 30% of their sales from coal and instead will grow its wind and solar portfolio.[92]

A similar initiative, backed by the UN, has been set up by the 350.org, a group of financial environmental activists, that has convinced 600 institutions to sell stocks and bonds of fossil fuel companies.[120] The most notable divestment member is the Norwegian $900 billion sovereign wealth funds, which has sold off $8 billion in coal assets in its portfolio, impacting 122 companies including E.ON, RWE, Dong, Vattenfall, and Duke Energy.[93]

120 The website of this initiative is www.gofossilfree.org.

Chapter 6 – What Will the Energy Mix of the Future Be?

China has seen rapid economic growth and significant improvement in people's lives. However, this has taken a toll on the environment and resources. Having learned the lesson, China is vigorously making ecological endeavours to promote green, circular and low-carbon growth. China will, on the basis of technological and institutional innovation, adopt new policy measures to improve industrial mix, build low-carbon energy system, develop green building and low-carbon transportation, and build a nation-wide carbon emission trading market so as to foster a new pattern of modernization featuring harmony between man and nature.
– Chinese President Xi, 2015[1]

The storage battery is, in my opinion, a catchpenny, a sensation, a mechanism for swindling the public by stock companies. The storage battery is one of those peculiar things which appeals to the imagination, and no more perfect thing could be desired by stock swindlers than that very self-same thing. Just as soon as a man gets working on the secondary battery it brings out his latent capacity for lying.
– Thomas Edison in The Electrician, 1883

We need a massive amount of innovation in research and development on clean energy: new ways to stabilize the intermittent flows from wind and solar; cheaper, more efficient solar panels; better equipment for transmitting and managing energy; next-generation nuclear plants that are even safer than today's; and more.
– Bill Gates, 2015

Introduction

The availability of cheap electricity on a continuous basis is taken for granted in high-income countries. A blackout causes chaos within hours in our highly sophisticated economies. The United States suffered a blackout due to a software bug in 2003, and it affected 50 million people for two days, stopping all electricity, lighting, phone services, air traffic, and train transport in the affected regions. In New York City, many people had to sleep in their offices overnight, food waste was dumped on the streets, emergency services were interrupted, raw sewage overflowed on the beaches, and water was not available in high-rise buildings as electric pumps no longer functioned.[2],[3]

In many countries the population is still far from fortunate in not having continuous access, with about 2.8 billion people suffering from power outages on a weekly basis in countries such as Tajikistan, Ghana, and Bangladesh.[4]

The more interrupted the flow of electricity is, the more expensive the electricity, and the more challenging it is to set up any continuous mode of manufacturing. The majority of factories today operate on a continuous basis thanks to the automation of most processes combined with 24/7 electricity. Nearly all bulk chemicals such as ammonia, chlorine, and sulphuric acid are produced in this continuous manner. That's why the manufacturing sector is practically absent in the entirety of sub-Saharan Africa. Restrictions are so severe that no economy with frequent electricity interruptions has been able to expand and grow beyond a size of $10,000 in GDP per person.

Equally, the flow of goods is dependent on 24/7 availability of crude oil today for trucking and shipping. The just-in-time delivery to supermarkets and stores relies on a well-oiled supply chain. Three days of trucker strikes in 2008 in Spain quickly led to empty supermarket shelves, panic buying of petrol, the shutdown of car factories, and cancellation of public flights and ferries.

Since the world is transitioning towards more fluctuating sources of solar and wind energy, maintaining our 24/7 mode for electricity is one of the biggest challenges for economies. The end of cheap oil and increasing demand for electric transport makes the challenge even bigger. The key issue for a clean energy future is: how can we manage our economies without fossil fuels?

53. Why is the electricity mix based on coal and gas?

Since the late nineteenth century, coal has been a dominant source for electricity generation. Natural gas entered the energy mix of many countries in the 1920s, and since the 1960s has become increasingly important. Today all the large economies of the world are dependent on these two fuel sources for the majority of their power. The exceptions are a few nuclear and hydro-powered nations. For example, France is over 70% dependent on nuclear power, and Brazil is over 85% dependent on hydro.

The main reason for coal and gas-powered economies across the world is their truly low cost, high energy content, and because we can easily store them and generate power when we want where we want. The highest-quality coals contain 26-33 MJ per kg, and natural gas contains 34-39 MJ per kg in energy value. Today the retrieval of 1,000 kilograms of coal and their conversion into electricity provides enough energy to power 10 refrigerators for an entire year, at a labor cost of less than a single hour.[5]

Natural gas has become the main source of cheap power in countries where large gas fields have been discovered, including the United States, Russia, northern Europe, and the Middle East. With nearly no processing or labor needed, its cost is extremely low. The price of natural gas in the US market fluctuated between $30 and $50 per 1,000 m³ in the twentieth century prior to the oil crises. It grew four-fold to $180 per 1,000 m3 during the oil crises, and stayed high until the rock bottom $50 per 1,000m3 price, as a consequence of the US shale revolution. Similar price trends can be observed in other countries (figure 16). Still today, a 1,000 m³ of natural gas, obtained with less than an hour of labor,

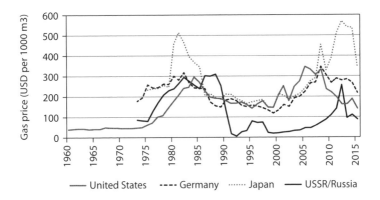

can supply heat for a single household for close to an entire year in colder climates.[6]

The rise of natural gas as a global source for power generation started in the 1960s. The first large-scale pipeline export of natural gas was established in the 1970s from Norway to northern Europe as well as Russia via Ukraine, and from Canada to the United States, and in the 1980s from northern Africa to southern Europe. Today these are still the biggest pipeline systems, through which nearly 70% of all pipeline-exported natural gas flows.[7]

Since the 1990s, cross-continent transport of natural gas has also become possible thanks to the liquefaction of natural gas. In this process, the gas is cooled down to -162 degrees Celsius (-260 degrees Fahrenheit) so that it becomes a liquid, after which it can be pumped onto a special storage vessel called a Liquified Natural Gas (LNG) tanker. As the tanker sails across the oceans it burns some of the gas to power the engines, and arrives at a facility where in turn the liquid is converted back into gas by gradual heating.

Once gaseous, the natural gas can be transported to a power plant in the country of destination. Today about 10% of produced natural gas is exported to other countries via LNG transport, and 20% is exported by pipeline.[7]

54.   Can shale gas fuel the US economy for decades to come?

Just like shale oil, the rise of shale gas, which is trapped in rocks that need to be fractured for gas to flow, has had a big impact on the US economy. Its success has been so enormous that an earlier expected peak in US gas production was reversed and coal consumption began to decline because many electricity producing companies have switched from coal to gas. Thanks to shale gas, North America has remained self-sufficient in gas despite growing regional demand, of which 46% is now supplied from US shale gas.[8] The latest US government estimate states that shale gas will be sufficiently abundant to fuel the US economy for several decades.[9][121]

The irony is that the success of the shale gas boom, just like shale oil, is leading to its (temporary) demise. The rapid growth in extraction has made investment in shale gas in the US increasingly economically unfavorable. The price of natural gas has declined by 65% since 2008, from $280 to below $100 per 1,000 m^3. The effect of the price decline, just as we have seen in oil shale, has been a rapid drop in drilling for new shale gas wells, from 1,400 rigs running at end of 2008, to around 1,000 active rigs in 2010, to less than 100 rigs mid-2016. Each of these can drill about 15 to 20 shale wells every year.[122]

121 The EIA estimates 5.6 trillion cubic metres (TCM) in the United States to be recoverable, relative to a 2015 extraction rate of 425 billion cubic metres. The US Geological Survey estimates another 12 TCM to be potentially technical recoverable from new evaluations and discoveries in its latest 2013 assessment, which will likely grow further as the resource base is re-evaluated such as the 2016 Mancos Shale assessment.[127]–[129]

122 And add about 0.3 to 1 million m^3 per day of production in the first year, depending on the productivity of the shale gas play.

Fig. 17. US natural gas prices, shale gas production, active gas drilling rigs, and drilled but uncompleted wells from January 2000 to June 2016.[7],[12]–[19]

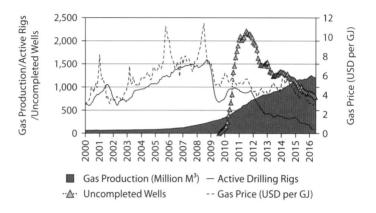

Shale gas wells decline rapidly from the start, at typically 60% to 70% within the first year, after which they produce at much lower rates for a decade or more.[10][123] The effect of stopping drilling new wells affects production abruptly within a year's time. US shale gas production only started falling in 2016, however, with a 4.5% drop. That this did not happen earlier can be explained because a number of wells were drilled several years before they were brought into production (figure 17).[124] Now that production is declining, gas prices have increased from a rock bottom $53 to $103 per 1,000 m³.[11] Only when prices shoot back up to pre-2010 levels will drilling be resumed at scale. In order

123 In contrast conventional gas fields typically ramp up production within a year, and that level is maintained for several years to over decades, after which a rapid decline sets in.
124 Due to a lack of crews that could bring wells into production, in combination with limited pipeline capacity to transport the gas, and for financial hedging reasons.[130]

to sustain long-term production, prices will likely need to be maintained above \$150 per 1,000 m^3 in the United States. Several decades of US shale gas supply are there, but will come at a substantially higher price.

## 55.	What about shale plays in the rest of the world?

So far there has been a notable lack of success in other countries in efforts to exploit shale gas, which has not been due to a lack of effort. Several companies drilled for shale gas in Europe, including in Poland, the UK, Denmark, Romania, and Lithuania. The biggest effort in Poland only led to 25 wells out of 72 wells with gas by the end of 2015, none of which had enough gas flow to be profitable.[20] All international companies have now abandoned their activities in Poland.

The next best promise in Europe was the UK. The company Cuadrilla carried out test drills in 2011 in the Lancashire region, where shale plays are relatively close to the surface, but this triggered two small earthquakes, and a moratorium on further fracturing was put in place by the government. Now that it has been lifted, companies could in theory start drilling again, but due to large public opposition from local residents over traffic, noise, environmental and visual impacts, local councils in the UK have so far halted any efforts to drill.[20]

The challenges in exploiting shale gas beyond the United States are related primarily to cost and scale as influenced by geology. Dominant US shale plays have a number of highly favorable characteristics. Despite the economic favorability—with gas prices of $236 per 1,000 m^3 in Europe—it is unlikely that any scale of shale gas production will be reached anytime soon for technical plus political reasons. Since only about 100 drilling rigs in total are in existence in all of the 28 EU countries, a large investment in infrastructure would be required, in addition to building experience in spudding and fracturing shales. The main countries where shale gas production for that reason may take off in the near-term future are not in Europe, but include China and Russia as both have a large drilling rig manufacturing industry, and already operate 1,500 and 850 drilling rigs, respectively.

Russia has not yet focused on shale gas development, as its conventional gas resources are much cheaper to produce at present.[21] Totally opposite are the Chinese, whose government set a target for 6.5 billion cubic meters (BCM) of shale gas production in 2015 and 60 to 100 BCM in 2020.[108] That would equate to about 25% of present US shale gas production. The 2015 Chinese target fell short, however, despite financially supportive policies. The only success so far has been the Fulen shale gas field in Chongqing province which yielded 2.6 BCM in the first half of 2016, with an anticipated 10 BCM of output by 2020.[22],[23]

The challenge is that most of China's shale resources are extremely difficult to extract, often deposited at depths of 5 kilometers and encased in hard, thick formations.[24] In the promising Sichuan province a large number of technically recoverable shale resources have been evaluated, yet the cost to get it out of the ground is at present \$430 per 1,000 m^3, at 23% and 252% above Chinese residential and industrial gas prices, respectively, as set by the government.[25] Oil company BP has not yet been deterred, and in September 2016 agreed with China's National Petroleum Corporation (CNPC) to develop shale gas in the Sichuan Basin, in a 400-square-mile area, following an earlier deal that had already been struck.[26]

In summary, in the near-term future, shale gas expansion at large scale will most likely remain limited to the US for the reasons of cost and infrastructure requirements, as well as policy and environmental concerns in some regions. While sufficient resources exist outside the US, it may be a decade or more before the success at the scale of US is replicated in another region, with China as a likely candidate given its aggressive policies. It will be several decades before global exploitation can take off, if it does at all.

Surprisingly, another source of unconventional gas, tight gas sands, holds a greater promise in terms of resource potential and is already extracted commercially in a few locations including the US, Canada, Indonesia, France, Argentina, Russia, China, and Venezuela. The exploitation of tight gas began in the United States in the 1970s. After a temporary lapse in activity in the aftermath of the oil crises, this source has grown to 15% of US natural gas production in 2015.[27],[28] Globally about 7% or 252 BCM of all gas produced comes from tight gas sands, nearly all in the US, Canada, Russia, and China.[29]

Total potential extractable tight gas resources would in theory be sufficient gas for the entire twenty-first century (table 6).[109] Tight gas sands are produced with the same hydraulic fracturing approach as shale gas, but only 6% to 10% of a tight gas reservoir can be extracted because gas migration in the sands is the hardest of all gas deposits (the gas is tightly packed in the sand, and there are few naturally occuring fractures through which gas can migrate).[36] Production rates are lower than shale wells, but can on the other hand be maintained at these lower rates for a longer time.[37][125]

Also many tight sand reservoirs contain water in addition to gas which has to be pumped out at high cost.[38] Some researchers are experimenting with underground heating to

125 This is because production relies less on breaking open natural fractures that provide rapid initial rates but deplete quickly.

'boil out the water' at 900 degrees Celsius.[39] The geological conditions of tight gas plays make their gas costly to extract in many cases. In the US break-even production costs have been estimated at $135 per 1,000 m^3 for core productive regions up to $360 for non-core low-quality regions of tight gas plays, by the analyst firm IHS Energy.[40] While these are normal levels for gas markets in EU and Asia, they give another indication that US gas prices need to head higher for unconventional gas production to remain profitable.[126]

The development of tight gas in China has been instrumental in the country's gas production growth, with about 20% of its 2015 gas output of 138 BCM coming from this unconventional resource type.[41] Production scenarios by the China University of Petroleum in Beijing show that this would be sufficient for production growth to 100–150 BCM per year by 2040 which can be sustained for several decades before decline sets in.[42] In the United States the US Energy Information Administration anticipates tight gas supply to continue its slow growth path to about 180 BCM by 2040 from 126 BCM today.[43] In Canada, tight gas already represents 53% of total gas output in 2015 at 82 BCM produced. The Canadian National Energy Board forecasts it to become even more important, with anticipated growth from 82 BCM in 2015, to 132 by 2030, and 141 by 2040, resulting in 76% of the country's gas production to be sourced from tight gas sands.[44]

---

126 In comparison in Egypt Shell has agreed upon a contract price of 200 USD per 1000 m3 with the Egyptian General Petroleum Corporation to develop tight gas from the Apollonia basin.[131] And in Oman, BP has signed a 30 year deal to develop the Khazzan tight gas project for 16 billion USD, which based on cumulative production and BPs project share amounts to approximately 190 USD per 1000 m3 for this tight gas source.[132],[133]

56.   Can cleaner natural gas replace coal?

Since the discovery of natural gas, the end of coal has been announced numerous times. Despite these forecasts, 'King Coal' keeps coming back. Electricity from coal in the last 15 years doubled globally, mainly due to a vast Chinese expansion. Between 2000 and 2015, China built 80% of all new coal capacity to generate electricity globally, at a staggering 850 power plants that can generate 775 GW of power, as much as the total capacity of both gas and coal electricity generation in the United States.[45],[46]

Surprisingly, amidst the Chinese coal explosion, several experts in the energy news headlines touted that the world was entering a golden age of gas. The Big Oil companies launched massive media campaigns touting that natural gas is the future of energy. In the UK Statoil launched a multimillion-dollar campaign 'Fueling the Future' that spearheaded natural gas. On the wave of the US natural gas boom, ExxonMobil sent the message that the company produced more electricity from natural gas than ever before as part of its 'Energy Lives Here' campaign. And Total spread the message that it is 'Committed to Better Energy' as a global campaign, including its commitment to natural gas. World leaders of the G20 group of countries have even caught onto the dash for gas in hosting a 'G20 Gas Day' at their summit in June 2016.

The OECD's International Energy Agency was the initiator of this upbeat picture for gas in a special 2011 report, 'Are We Entering the Golden Age of Gas?' Its set of conclusions forecast natural gas growth of 50% by 2035, reaching 5.1 trillion cubic metres, with growth mainly from non-OECD countries, especially the Middle East and China.[47] If realized, this development would result in natural gas overtaking coal's present dominant role in electricity and heating supply by 2030.

The most striking conclusion by the IEA is that China would become one of the largest gas producers in the world and consume as much as all European Union countries together today. The country is still heavily reliant on coal with only a 6.2% share in the energy mix for natural gas.[48] So far the country has failed to come close to the IEA's expectations, and fell 15% short of its own 150 billion cubic meter (BCM) production target for 2015. China has instead begun importing a substantial 63 BCM, which is expected to grow according to the China National Petroleum Corporation (CNPC) to up to 270 BCM by 2030.[51],[52],[49],[50]

Where would such continued growth in natural gas, equivalent to three times current Russian natural gas production, come from? And will it overtake coal? The IEA thinks the world is comfortably positioned to sustain this growth well beyond 2035 from virtually all regions. The largest centers of growth would be the Middle East, Russia, North America, Caspian Sea countries, China, and Africa. The only 'losers' are expected to be the North Sea countries in Europe where supplies are anticipated by the IEA to continue the decline that started in the 1990s. Of all this gas growth, about 40% would come from unconventional shale gas, tight gas, and others.

So far we are on track, as gas demand is increasing at about 2% globally per year, in line with the golden age scenario, and coal is losing its growth impetus. About 100 new natural gas power plants are built every year with a joint 58 Gigawatt capacity, and natural gas heat use is equally growing substantially. Yet to sustain such growth in the long run, the rise of unconventional gas is key, far beyond the shale gas boom in the United States. Total current global conventional gas reserves stand at 193 trillion cubic meters (TCM), with another 160 TCM potentially added over many decades from discoveries and enhanced recovery techniques (table 6).[127] In contrast global natural gas production in 2015 was 3.5 TCM, thus reserves are at best 55 years of supply.

We know the resources for unconventional gas are available, with a lower range of estimates calling for 167 BCM recoverable unconventional gas, and an upper end of the range at 1,356 BCM (table 6). Potentially unconventional gas is thus sufficient to last far beyond the twenty-first century, but we cannot count on it yet to last that long, and it will likely be a costly affair. According to the IEA the 'golden age of gas' expansion comes at a price tag of $8 trillion in investment needs between 2010 and 2035, as estimated from extraction cost and infrastructure evaluations. To achieve commercial extraction of enough resources, a tripling of US gas prices and growth by 50% in other markets are needed, according to the IEA in its gas scenario.[128]

127 Values based on resource assessments by the US Geological Survey, German BGR, World Energy Council, and Schlumberger plus Texas and Houston University researchers.
128 To an estimated to 283, 386, and 456 USD per 1000 m³ in the U.S. Europe and Japan, respectively as calculated by the IEA and relative to these countries 2015 price levels of 93, 236, and 368 USD.

Table 6. Estimated Global Gas Resources and Reserves in TCM. Sources:[30]–[35]

Resource Category	Already Produced	Discovered Gas left in Place	Potential Proven Reserves
Conventional gas	108	318	193
Yet to find conventional gas estimate	-	344	133
Potential resource-to-reserve EGR**	-	50	27
Conventional gas	108	712	353
		Resource Estimate	Technically recoverable
Shale gas	n/a	455 - 1422	115 – 204
Coalbed gas	n/a	115 – 256	14 – 118
Tight gas sands	n/a	209 - 2044	38 – 1034
Gas hydrates	n/a	184	0
Unconventional gas	n/a	963 - 3906	167 - 1356
TOTAL Low to high estimate	112	1675 - 4618	520 - 1709

** Enhanced Gas Recovery (EGR) based additions from discovered gas left in place to technically recoverable

57.   Do we need to reduce world coal use?

Coal has always been assessed to be the most abundant fossil fuel resource, with several hundred to over a thousand years of supply. It is clear that the world is awash with coal deposits. Recent estimates for extractable higher-quality coals range from 688 to 700 billion tons versus a close to 7 billion-ton production level in 2015.[34],[35] Moreover, these estimates by the World Energy Council and the German geological survey BGR, only account for coal in the first kilometer to the surface in onshore areas.

Its abundance means that problems with coal supply will only appear in a few regions. Coal mining in the US states of West Virginia and Virginia, which still account for 13% of US coal output, has been dropping since 1990 and 2008, respectively.[129] This is compensated by growth in coal mining in regions in the US west of the Mississippi River where coal is much more abundant and cheap. The biggest contender for future depletion and price growth today is China. Coal extraction grew between 2000 and 2013 as never seen anywhere before, from 30% to 48% of world coal output, or in absolute terms from 1.4 to 4.0 billion tons per year.[7] Yet even at such high production levels, the country has coal reserves available for four decades, plus far more costly-to-produce deeper-lying coal.[53]

---

129 Production has dropped in Virginia from 42.3 to 12.6 million tonnes between 1990 and 2015, and in West-Virginia from 143 to 95 million tonnes between 2008 and 2015.[134]–[136] Coal mining costs have grown as labour productivity dropped from around 3000 to 2400 kg per hour of work from 2007 to 2014 in West-Virginia, and 2800 to 1700 kilogram per hour of work in Virginia between the late 1990s to today.[137]–[141]. At such higher costs, in combination with low coal prices, mines have been outcompeted by cheaper supplies. Even federal state tax credits have not helped to stem the decline.[142],[143]

Temporary problems in supply can appear, as the doubling of traded coal prices of 2009 and 2012 show. These were the result of China becoming a net coal importer despite its enormous production. No shortages occurred, and the price has again dropped substantially as Indonesia managed to increase its coal exports to China by no less than 50% in just three years, while Chinese import growth stabilized.

The bottom line is that the future of coal is not really dependent on what is in the ground, but more on what price we are willing to pay (and thus the number of excavators and the amount of fuel we can throw at getting coal). The extent to which coal cost will rise is anyone's guess. If the largest US coal field, Gillette, is a representative example, we can be optimistic, as this field will yield three to five times more coal at a doubling to tripling of the US coal price.[54] Also, deeper-lying coal can be extracted by a process of controlled burning in the coal cavity underground via a technique called underground coal gasification (UCG).[130] It was developed in the 1960s by the Soviet Union, but it has only gained a foothold in smaller projects in Australia, South Africa, and Canada. At today's price for mined coal, the process is not commercially interesting.

130 Underground coal gasification induces a controlled fire into a coal seam by supplying oxygen, and the resulting gaseous mixture from combustion is pumped out to the surface which includes methane, hydrogen, carbon- dioxide and monoxide.

58.   Will coal be phased out because of pollution?

Since coal incineration emits close to 50% of all fossil fuel-related carbon emissions, it has always been the first target in climate change discussions.[55] The first global agreement made by more developed countries to reduce emissions was the 1999 Kyoto Protocol. In this internationally binding agreement, 37 countries agreed to reduce their carbon emissions by 2012 to below the 1990 level, including the EU countries, Japan, and Russia. They succeeded in the reduction goal but primarily as a consequence of the financial crisis of 2008 and low economic growth, and only secondly due to government policies. In the same time period, at a global level, emissions of carbon dioxide increased by 52% from 1990 to 2012, due to growth in developing countries which were not part of the agreement.[55]

Coal use in the 37 Kyoto countries was reduced 9% by 2012, and 17% by 2015. But this was solely because of a drop in US coal use, coalesced by market forces and state level policies, as coal is increasingly being replaced by natural gas in the country. In the other 36 countries, coal has largely stayed at its earlier levels. Even in Germany, one of the most aggressive countries when it comes to solar and wind electricity generation, coal hardly declined at all, and still provided 44% of electricity in 2015 versus 45% in 1999.[56] The only country that may in the short term succeed in phasing out coal is the UK, a country which based on the Climate Change Act of 2006 instituted an 80% emissions reduction by 2050 in combination with a legally binding 2% annual reduction. As part of its framework, it is phasing out all coal power stations and replacing them with natural gas and imported biomass pellet fuel from the US.

The failure to reduce coal use in nearly all high-income countries begs the question if we can easily do without cheap coal. If as of today no new coal power plants are built, coal use for electricity generation would end by 2065, and 50% fewer power plants would be operating in the 2040s (figure 18).[131]

Low- to middle-income countries with a lot of coal and limited natural gas resources will have an especially hard time phasing out coal. Investment costs to build a 1 GW coal power station in Asia was estimated at $200 million, versus $1.1 billion for a natural gas plant of equal size.[57] Also fuel prices are vastly cheaper, with locally mined coal in India and China costing less than $2 per GJ of energy.[132] To substitute natural gas would require imports, which via pipeline from Russia so far costs between $8.50 and $10.50 per GJ.[58] LNG ship-based imports from countries like Australia, Qatar, Malaysia, and Nigeria in the last 10 years have fluctuated between $4.80 and $14.20 per GJ.[59],[60]–[62] And imported coal is far cheaper, such as from Indonesia to China which fluctuated between $2.00 and $7.60 per GJ in recent years.[63],[64]

The main reasons to phase out coal today are thus not due to cost or availability but climate change and to reduce local human health costs.[133] Global coal generation capacity today is close to 1,800 GW. Because many power plants are nearing their 40 to 50 year retirement age, and few countries are still expanding coal-based power generation, the current

131 This assumes a coal power plant life-time of 50 years.

132 The energy content of 40 to 60 kilograms of Chinese or Indian coal is about one gigajoule as used in the cost calculation.

133 In China and India an estimated 1 and 0.6 million people die prematurely every year due to outdoor air pollution, respectively, mainly from vehicle exhaust and coal burning.[144]

Fig. 18. Global development of existing coal power plant capacity based on an (optimistic) retirement age of 50 years.[65]

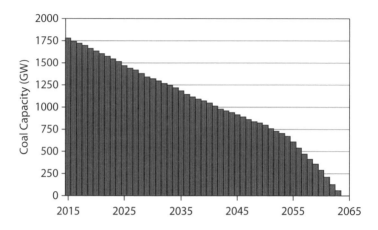

expectation is that coal supply will likely stay stable for now.[134] The two countries that could change this picture are China and India, with 15% and 50% of all new coal power plants under construction, respectively.[135]

China has since 2005 worked heavily on reducing the energy and carbon intensity of its economy and invested massively in clean energy sources. Today China is globally the largest investor in clean energy, and the biggest

134 Globally a total 210 GW of new coal power plants are under construction, and another 320 GW globally is either approved or in the bidding process. A total of 330 GW coal power plants will stop to operate by 2025 due to old age, and another 290 GW by 2035 (based on a 50 year retirement age).[65],[145],[146]

135 The US also could play a role, as 39% of its electricity still comes from coal. The Obama administration tried to advance the phasing out of coal via the Clean Power Plan, which was proposed to Congress in 2015. The act sought to reduce carbon dioxide emissions by 32% from power plants relative to 2005 levels.[147] It would force older power plants out of action, substituting newer coal fuelled with natural gas power plants, and increase solar and wind energy. A US Supreme Court ruling has so far barred the plan from being implemented.

in producing electricity from solar and wind. It is also experimenting with several regional carbon dioxide emissions trading systems.[66],[67] So far, Chinese policies have resulted in a stabilization of coal usage at present rates, and China is showing signs that coal use is declining.[68] Despite many more coal power plants being built, their use is declining. In 2015 China's coal power plants only operated 50% of the time in 2015 for 4,329 hours, versus 57% in 2013.[69] Far too many were built in low-demand northern coal regions with too few power lines in place to transport surplus electricity to high-demand regions until about 2018. Chinese coal production likewise declined since 2013, with a 20% drop in production to 3.2 billion tons in 2016.[70] It looks like China does not need more coal power.[136] The central government is now slowing down coal expansion, with a moratorium on new power station approvals until 2018 and a maximum increase of 20% in coal generation capacity by 2020 from existing approvals (to 1,050 GW).[71]

India is on a similar course of coal slowdown and aggressive pursuit of renewables. The government made the smart move to put a tax on coal in 2010 to support renewable energy development. The tax has been doubled four times since, and is now $0.30 per GJ of coal consumed.[72] The revenue is much needed as the country is investing heavily to reach its ambitious targets for 100 GW of solar electricity by 2022, and 75 GW of wind- and biomass-based generation.[73] At the same time its power estimates for 2013 turned out to be about 20% too optimistic, and about 50 GW less coal capacity is needed in the time frame to 2022.[74] The government has announced it won't need any new

136 China is currently constructing 29 GW and has 116 GW in coal power plants approved or in the bidding process.

power plants for the next three years, and due to lackluster regional interest, 16 GW of large coal power plants have been dropped that were already approved earlier.[75],[76]

If India and China continue on this path, coal is likely to become a fuel of the past for electricity production. Goldman Sachs analysts Christian Lelong and Amber Cai may turn out to be right, that 'peak coal is coming sooner than expected', as outlined in an investor presentation in 2015.[77]

59. What role will nuclear energy play in the future?

Since the Fukushima meltdown in 2011, political support among the world's nuclear superpowers has gone in different directions. Germany decided to shut down all of its 17 nuclear power plants. The Japanese government decided to mothball all 53 of its reactors, which led to vast increases in LNG tanker imports as a replacement. Japan now seeks to restart half its reactors by 2017, after additional safety implementation.[78]

France has adopted a new energy law to reduce its share of nuclear to 50% from the current 75% by 2025, and Canada has deferred any expansions but will maintain its current nuclear base.[79]

But the world's use of nuclear power will potentially rise because over 168 new nuclear reactors are in an advanced planning phase, while 400 are in operation worldwide. Globally some 59 reactors with a joint capacity of 62 GW are now under construction, primarily in China (20), Russia (7), India (5), the US (4), and the UAE (4). Today China is pushing nuclear energy the hardest with 34 reactors in operation, 20 under construction, and another 42 in advanced planning.[80] Even Saudi Arabia aims to construct dozens of nuclear reactors. In the UK, the British government just approved the start of the first nuclear reactor in a very long time, the Hinkley Point nuclear plant, which will cost $24 billion to build. Because of these developments uranium demand could grow by 2%-3% per year, following total capacity growth from 390 GWe in 2016 to 600 GWe by 2030.

The big unknown on the downside is whether nuclear power plants will be allowed to operate beyond their initial design age of 40 years. In the US so far, 74 out of 99 operating reactors have received a lifetime extension from 40 to 60 years. Extensions do come at a significant cost, however, to upgrade reactors to a desired substantial safety level.[81] If extensions are not granted, over 200 reactors older than 30 years with a joint capacity of 160 GW would be shut down before 2035, which would mean that nuclear power generation would be about the same in 20 years as it is today.

Another big potential downside is the ability of nuclear to operate successfully commercially in today's power markets. In the United States, 14 reactors, which had received lifetime extensions, are being closed prematurely because wholesale power prices have dropped below $30 per MWh result in large losses.[82] About 55% of nuclear power plants are estimated to be in the red figures in the next three years, as nuclear operational costs in the US range from $33 to $44 per MWh.[137] The remaining plants can still make a profit because of additional reserve capacity payments, to produce power if needed in the future to stabilize electricity grids.[83] In the UK the key reason Chinese and French investors agreed to put money into Hinkley Point was a government-guaranteed power sales price of $120 per MWh. If wholesale power prices in the UK stay around today's level,[138] the price guarantee entails a government subsidy of $40 billion over its lifetime.[84],[85]

The future of nuclear hangs in the balance of mainly political decisions. In the long term, the picture for nuclear could still be flipped if old nuclear designs developed in the 1950s are proven to work. These include 'breeder' or 'fast' sodium-cooled reactors, molten salt reactors using thorium

137   The biggest causes of the low wholesale power price in the U.S. are low natural gas prices, and the guaranteed sale of wind and solar whose marginal cost is zero, which drives down power prices.

138   Between 50-60 USD per MWh.

instead of uranium, and gas-cooled pebble-bed reactors. Over the decades, despite a number of experiments, none of these has gained any prominence due to small accidents, high capital costs, and concerns over safety and proliferation. But novel developments, especially in China, Russia, and India, are potentially leading to a renaissance in novel nuclear fission.

The key concerns about current pressurized-water nuclear reactors include the radioactive waste generated, and their inefficient use of only 1% of uranium in a given fuel load. To tackle this issue a fast sodium-cooled reactor called BN-800 with 880 MW capacity was opened in Russia in 2015, the first since smaller test reactors in the 1980s. The reactor runs on uranium-plutonium mix, thereby reducing the Russian weapons-grade plutonium stockpile. This type of reactor can also run in 'breeder' or 'fast' mode to utilize up to 100 times more of the uranium in the fuel than conventional reactors do, thus vastly extending the availability of uranium.[139] Two similar Russian reactors will soon be operational in China by 2019.[86],[87] Based on the current experience, the Russians anticipate that by 2020 a new improved design called BN-1200 with a capacity of 2.9 GW will be ready.[88]

Substantial progress has also been made in China with gas-cooled pebble-bed reactors.[140] Only two large 300+ MW reactors have ever gone into operation, in Germany and the

139 Breeder reactors also produce plutonium which on its own can be used as a fuel (or in nuclear weapons).

140 Gas-cooled pebble bed reactors work by flowing helium gas at the reactor top, which is heating through the nuclear reaction to 750 degrees Celsius (1380 degrees Fahrenheit), which can then be used to drive a steam-turbine for electricity generation. The nuclear reaction is contained in hundreds of thousands of small uranium pebbles flowing and bumping within the helium. These are loaded continuously as blown by the helium gas at the top; they stay in the reactor for

United States in the 1970s–1980s, both shut down for reasons of cost, operational difficulties, and lack of political commitment. Two new 105 MW pebble-bed reactors deemed to be immune to nuclear meltdown have been constructed in China and are currently undergoing testing, before reaching full operation in November 2017.[89][141] If successful, China aims to develop a standardized 600 MW size reactor based on the existing design.

Developments in nuclear fusion are much slower than in advanced fission designs. In 2006 the EU, Japan, India, China, Russia, South Korea, and the United States published news that they would fund the first practical fusion reactor together. The reactor would initially be built in Cadarache in France for $5 billion within 10 years, but that proved to be incredibly optimistic. Total costs have risen to $16.6 billion and the start date for first plasma has shifted to 2026.[90],[91]

This is not a commercial project, and the aim is to generate 10 years of experience before a 200 MW demonstration reactor would be built that initially would start operating

fission; and come out of the bottom alongside the controlled helium flow, thus reducing reactor maintenance shutdowns.

141 The gas-cooled pebble bed reactor is immune to a nuclear meltdown because the uranium pebbles are coated with graphite which does not melt or burn, as graphite literally does not have a melting point – it will only turn into a gas directly at the extremely high temperature of 3456 degrees Celsius (6254 degrees Fahrenheit). Moreover, the helium gas in the reactor is inert and cannot explode; the reaction in case of loss of cooling slowly shuts itself down due to a large negative reactivity feedback (the higher the temperature the slower the nuclear reaction), and faster passive heat removal as per the reactor design.[148]–[152] All these safety features do not mean that there is no risk of accidents, however, as in the German 300 MW reactor on May 4 1986, only a few days after the Chernobyl nuclear accident, a pebble got stuck in the reactor helium feeding pipe. In an attempt of the operators to unlodge it, a small dose of radioactive dust was released within a 2 km radius of the reactor, destroying its touted accident free reputation.[153]

in 2040 (now likely delayed until at least 2050).[92] If commercial fusion happens, it will not be until the second part of this century. Nuclear fusion has been a dream that is at least 30 years away for over 30 years now.

60. Can we rely on wind and solar for our electricity needs?

Since there is no solar energy during the night, and wind output comes with large fluctuations from zero to 100%, either backup or storage of electricity is needed for 24/7 grid operation. Storage solutions need to work for the short- (minute to hour), medium- (day to day), and long-term seasonal storage requirements (week to month). Until the storage issue is solved, large amounts of renewable sources on grids will need to be complemented with fossil fuels, especially natural gas, to make sure electricity is available 24/7.

If we look at windmills on land in Germany, the United States, and China, a wind turbine in the last five years provided electricity on average about 19%, 29%, and 15% of the time, respectively.[142] Even at the windiest places in these countries, windmills operate about 35%-40% of all the hours in the year.[143] And wind can jump from very low amounts of electricity to high in a matter of minutes to hours.[93] Similarly, solar power generates power on average between 9 a.m. and 5 p.m. and does not deliver for two-thirds of the day. In contrast, baseload coal, natural gas, and nuclear power sources can provide electricity 75% to 85% of the year.[144]

142 The value for China is artificially low, because many wind-farms were standing still in recent years because of a lack of long-distance transmission possibilities from North to East China.

143 Values obtained from measurements at different locations in Germany using the Renewables Ninja App developed by researchers at ETH Zurich and imperial College London.

144 In practice coal, gas, and oil provide power about 60%, 40%, and 27% of the time during the year globally on average. Natural gas is used in many places to balance out all other power sources, and oil power plants as a back-up for power outages. Only coal in all cases is used as a baseload continuous power source.

The current way to deal with these fluctuations is by creating a lot of flexibility in electricity generation, mainly by using flexible natural gas power plants or hydroelectricity that can easily and quickly increase or decrease their electricity output.[94] Denmark obtained 42% of its electricity from wind in 2015 which is balanced using imported hydroelectricity from Norway.[95] In Germany over 20% of electricity—provided by wind and solar last year—is backed up by large numbers of natural gas power plants.[96],[97] The increase in renewables in Germany has however had a severe economic effect on companies with fossil fuel power generation installations. As a consequence of increasing renewables on the grid, wholesale kWh prices dropped from $0.09 to below $0.03 cents per kWh between 2008 and 2016.[98],[99] Companies with many fossil fuel plants in Germany, like RWE and E.ON, are now making large losses, with their stock prices halved since 2008.[100],[101][145,146]

Finally, renewables themselves have been made flexible as well. The grid operators in charge of making sure

145 The electricity regulators of Germany are trying to fix this by using a so-called capacity market, where payments are made to natural gas power plants to stay in business, even for times when they do not operate. This so called 'capacity market' will start in 2017 where about 7% of natural gas power plants purely stand in 'wet reserve', which can be brought into production when needed to match wind and solar. The total cost of this solution is estimated to be 145 to 290 million US dollars per year.

146 Earlier already a so-called 'reserve market' was created. This works by financially compensating companies to within minutes to an hour increase or decrease their electricity output flexible natural gas power when there is suddenly less or more wind or solar. The natural gas power plants do operate, but not at full capacity to allow for increases as well as decreases in generated electricity. In Germany the available in 15 second reserve is about 0.58 GW, the available in 5 minutes reserve 4 GW, and the available in 15 minute reserve 5.2 GW, with about 50:50 of the 5 and 15 minute reserve downside and upside control.[154]–[156]

electricity is there 24/7 have the legal right to reduce or fully halt the electricity output of windmills to keep grids stable when the wind blows too much.[102],[103][147] Together with these measures and high-quality weather forecasts, Germany can deal with up to 30% to 40% electricity from sun and wind. When penetration grows beyond 60%, as per the German government's 2040 target, daily electricity storage like batteries becomes necessary.[94]

Various short- to medium-term battery storage solutions are progressing. The largest push for battery electricity storage is taking place in California. A mandate from the California Public Utilities Commission is pushing utility providers to procure 1.32 GW of small- to large-scale electricity storage systems by 2020, despite their higher costs, which equates to about 1% of installed power capacity.[104][148,149]

An entirely different technology for medium term storage is compressed air, where electricity drives a compressor that

147 With growing wind electricity in large parts of Germany, an increasing number of curtailments is taking place at 6 in 2010 versus over 1000 in 2015, estimated to have led to a 'lost' wind-power output costing 280 million US dollars.

148 One of the first facilities to be contracted in California is a lithium-ion battery plant by AES energy storage to provide power for 4 hours at 100 MW by January 2021.[157],[158] The AES Advancion solution is stated to cost 1000 USD per kW, or 100 million USD for a 100 MW facility, with its batteries provided by South-Korean firm LG Chem, at a round-trip efficiency of 90% and 6000 charging cycles.[159] Another contender in California are the novel zinc based batteries from EOS energy, which has substantial cost advantages over lithium, yet they are not yet as durable but the company states it has solved this problem.[160]

149 The key problem is cost, since batteries are too expensive versus their fossil fuel competitor, natural gas 'peaker' plants, which provides similar services by burning gas mainly at times of high electricity demand (and low renewable output). A peaker plant in Germany costs about 66 million USD for a 100 MW facility, and is still at least 30% cheaper then batteries, even when ignoring the fact that batteries do not generate electricity, whilst natural gas plants do. A natural gas plant also has a longer lifetime of up to 40 years, whereas lithium-ion batteries so far come with a warranty of 10 years and 10,000 charging cycles.

puts normal air in an underground cavern or tank under extremely high pressures, and then uses the decompression as a means to generate power. Also thermal storage technology can be utilised, either by direct concentration of the sun on a medium like salt that is melted, or graphite blocks or packed basalt rocks that are heated to 550 degrees Celsius, or by using electric heating. The heat in the medium can be transferred at a later point in time to water to produce steam to drive a turbine for electricity generation.[150] These technologies are still far from perfect with relative large losses of energy in a few days.

Solutions for long-term storage of multiple weeks to months are an even bigger headache. Batteries are too costly and lose power too quickly. The most mature systems with about 500 GW in place are called pumped-hydro plants, so far the only commercial long-term storage options, but limited to places with elevated water bodies.[105] They operate by pumping large volumes of water in a low power-demand period from a low to higher elevation reservoir, and vice versa when power is needed. A more flexible future long-term storage technology with potential is power-to-gas-to-power, where excess solar or wind power in times of low demand is used to generate hydrogen or methane, which can be used to generate power again when it is required at any desirable time.

150 The German Energy Technology and Electronics company Siemens is currently pursuing a pilot thermal storage plant of 30 MW to be built by 2019.[161]

61. Should we all share energy through a worldwide grid?

The importance of geography and grid connections became painfully clear in China in recent years. Too many coal and wind power stations have been built in China's northern provinces without the grid capacity to connect them to the high-consumption provinces in the east.[151] A five-year spending program of $350 billion, 2016–2020, aims to resolve this by building another 20 high-voltage direct current (HVDC) lines from north to south and east of over 800 kilometers each.[106]

A recent key breakthrough to transport electricity flexibly is opening up many new possibilities for massive continental-scale 'supergrids.'[152] These can bring benefits of the seasonal complementary nature of wind and sun. It has been established that—at least in Europe—there are a lot of cloudy, windy days in winter and clear, still days in summer due to which the seasonal storage issue would largely disappear by combining wind and solar power (figures 19a & 19b).

151 The over-supply is so large that in the entire country 15% of potential wind-power, 30% of coal-power, and 31% of solar-PV power went unused in 2015.[69],[162]–[165]

152 A 800 kilometer long 400 kV AC overhead cable has a transmission loss of 9.4%, whilst a 800 kV HVDC line has a 2.6% loss.[166] The problem with HVDC lines until recently ewas that they were point-to-point and are not flexible multi-directional nodes, such that power can be re-directed as required to where it is needed, much like a car can move flexibly across a road-network. The common Alternate Current (AC) grid that we have today everywhere has this flexibility, The grid company ABB in 2012 fixed this 100 year old problem for HVDC by invention of a new grid 'breaker' for HVDC that can do the flexible re-direction at less than 0.01% loss within 5 milliseconds.[167]

Fig. 19. a) solar power generation in Germany in 2014 based on 40 GW installed capacity, b) wind power generation in Germany in 2014 based on 45 GW of installed capacity, c) example of connecting 240 GW of Wind and 90 GW of solar between Germany, Italy, Spain, Greece, Croatia, the UK, and Sweden in January 2014. Source of solar and wind data: http://www.renewables.ninja.

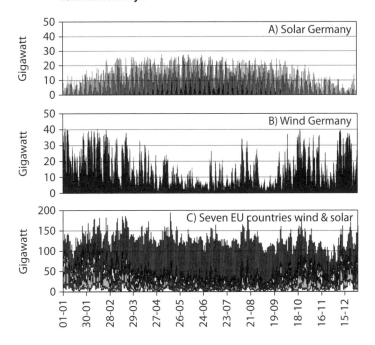

Interconnection over large geographic distances can also connect different wind regions to each other, for example, the North Sea, Baltic Sea, and Mediterranean Sea countries. The rationale is that connecting them would smooth out a lot of the day-to-day fluctuation between these countries. However, it is not an end-all solution, as can be seen in figure 19c, where in a theoretical example 240 GW of wind and 90 GW of solar capacity for the year 2014 are pooled across seven EU-28 countries.[153] The key challenge of super-grids lies in the enormous amount of planning they require, especially when they go through built-up environments. In Europe for this reason HVDC cables have been built only in offshore areas. Germany is making its first steps towards a large onshore HVDC grid. Three high-voltage direct-current lines will be built in the next five years, connecting offshore and onshore wind farms in the north to large electricity users in the German south.[154]

China is building its own supergrid. The country already has 20 such cables at 500–800 kV; the longest is 2,210 km long. It has also signed an agreement with the Russian, Japanese, and South Korean national grid companies to explore building a pan-Asian supergrid. And Chinese ambitions don't stop there. Zhenya Liu, the chairman of China's State Grid Corporation (SGCC), dreams of building a global-scale grid system by the year 2050 called the Energy Internet to which all continents are connected.[155] He launched the first Global Energy Interconnection conference in March 2016 where hundreds of international companies and institutes were brought together to talk about the global supergrid.[107],[108] Technically-speaking, creating such

153 Fluctuations continue, with in close to 8% of day-time hours of the year 50% or less of the average electricity output produced, and in two separate days output drops to around 20% of the average for several hours. Sufficient renewable power generation within the 24 hour cycle all year long can be established, but not for continuous hourly delivery without large day to day fossil back-up or storage in place.

154 These 1 to 2 GW 1000+ kilometer long 'power corridors' are facing substantial public opposition, however, and are now to be built using underground HVDC links where possible at 3 to 10 times the cost.[166],[168]–[170]

155 Bringing together the rich wind-power regions of south Greenland to Canada and Europe via Iceland, connecting Europe with Asia and Africa via the Middle East for solar and wind, connecting Asia to North-America via Russia and Canada, and also build direct links from Asia to Australia.

massive power corridors is a massive challenge, let alone economically.[156] Yet it would allow the connection of solar power from longitudinally very distant places, so that locations where it is night become connected to places where it is day, and vice versa.

---

156 The company ABB has developed a 1100 kV HVDC transmission line for links up to 3000 kilometers, with a 10 GW transfer capacity.[171] Chinese company SGCC has developed similar technology, with the first of its kind now under construction in China from the North-west to the South. SGCC thinks that by improving such technologies, eventually it would be feasible transmit wind or solar power at 4 cents per kWh in transmission cost from Xinjiang west-province to Germany, via Kazakhstan, Russia, Belarus, Poland, and Germany.[106]

62.   Can 24/7 trucking and shipping be converted to electric?

The use of fossil fuels in shipping has grown rapidly since the first container ships were introduced in the 1950s. Today over 100,000 ships use 5 million barrels per day or 4.5% of world oil production, mainly to haul goods across the globe.[100] Similarly, over 75 million heavy trucks run on almost 11 million barrels per day or over 12% of all the oil produced.[101]–[103]

Ships and trucks use heavy fuel oil and diesel straight from the oil refinery. Because of their high energy content, vehicles can go far with a relatively small tank. A heavy truck can carry 20 tons of cargo over 1,000 kilometers on a 400-liter diesel fuel tank, and a cargo ship with 8,000 containers can travel 20,000 kilometers with its 3.5-million-liter fuel tank. Moving such big vehicles with electricity would require massive batteries. So electric shipping and trucking would need super-batteries for electrification to make sense.[157]

Only short-distance trips for small vehicle delivery in cities or for inland ship transport are viable today. The largest hybrid electric battery ship so far is the *Princesse Benedikte* ferry in Denmark, which has a 365 car and 1,140 passenger capacity. This 15,000 ton ship can be propelled on battery power for 30 minutes with a charging time of 30 minutes.[104]

Short-distance electric delivery vans and light-to-medium trucks with a range of 160 kilometers on one battery charge are sold in the United States and Europe, but their

---

157 A potential candidate is a Lithium-oxygen battery, as these can hold five times more energy than the Tesla electric lithium-ion batteries in theory.[179],[180]

vendors are struggling to make a profit.[105]–[107] Build Your Dreams (BYD), the Chinese Tesla, has launched a series of electric vans and trucks that could get traction in the Chinese market thanks to government subsidies.[108] Of the big car companies, only Mercedes-Benz has unveiled a medium-sized electric truck which is not yet for sale, and Elon Musk of Tesla hinted that it also intends to enter the short-distance trucking space.[109],[110]158

Replacing oil in long-distance goods transport will require another liquid fuel, either natural gas, hydrogen, or biofuels. Hydrogen when compressed to 700 bar pressure holds 3.6 times more energy content than oil products, while biofuels typically contain half the energy.[111] The secretive US company Nikola plans a heavy-duty, 1,200-mile range hydrogen truck in addition to a planned 50+ hydrogen station network, one in every US state. The cost to produce hydrogen from wind power, however, is over $200 per barrel.[112],[113] The cost of biofuel is also still too high, with current estimates for biomass-to-methanol fuel at $70 per barrel; together with the biofuel energy penalty, this results in an oil-price-equivalent price of $150 per barrel.[114]

158 Another 'old' new technology which is moving ahead are electric hybrid trolley or catenary trucks for repetitive short-distances using, such as from a port to a distribution centre. The system is being tested by Siemens in Sweden and California at present on several kilometer stretches of highway.[181],[182] The overhead technology from Siemens as well as Hitachi is already used cost effectively in copper and iron mines for uphill haulage in heavy dump trucks with 320 tonnes load capacity.[183]–[188] The system allows twice faster speeds then diesel fuel can offer and reduces energy use by 40% at the mine level. About 20% less haulage trucks are needed as a consequence, and the capital investment pays itself back in 2 to 4 years.

63.   So biofuels are not a big success?

The use of biofuels in road transport is the biggest alternative to oil to date. Biofuels are made out of organic materials like trees and plants and are divided into three types. The first generation is converted from sugars, fibers, oils, and fats—thus mainly food crops. The second generation is based on cellulose, the woody substance that gives strength to plants and trees. And the third generation comes about from lipids and fats from algae.

Only the first crop-based fuels are available at present with 1.4 million 'bio-barrels' of 159 liters consumed in 2015 in the form of bioethanol and biodiesel. The main country where biofuels are commercially used at large scale is Brazil. Since the oil crises in the 1970s, the country has initiated a successful sugarcane-to-ethanol program. In 2015 the country used half a million barrels of bioethanol per day, based on a 25%-27% gasoline and 7%-10% diesel blending ratio. Brazil uses about 1.4% or almost five million hectares of its vast agricultural land to grow biofuels.[109],[110]

Since 2005 the production of biofuels has grown substantially as the United States and European Union governments have followed Brazil's lead and mandated blending for gasoline and diesel fuels.[159] The US 2007 renewable fuels standard calls for 2.4 million barrels per day of biofuels blended into gasoline by 2022.[111] In 2015 the country produced 600,000 barrels per day solely from corn as a feedstock, which required 40% of all corn grown.[7],[112]

The EU-28 countries agreed to a similar directive in 2009 at a strict 10% blending-of-biofuels target by 2020

159 Today about 25 other countries also include blending mandates for biofuels but none produce as much as the US, Brazil, and the EU-28.[172]

that equates to about 750,000 barrels per day. So far, the EU-28 together consumed 270,000 barrels in 2015 primarily from sugar beets and rapeseed planted on 25 million out of 312 million hectares of agricultural land, showing the inefficiency of these crops relative to sugarcane in Brazil.[113],[114] Because of increasing popular opposition to food competition impacts, the EU position has shifted in recent years, however, and its biofuels target will be dropped after 2020.[115]

The policy overhaul in the EU is a sign that biofuels are, at least for now, running out of steam. Despite grandiose announcements during the high oil price period of 2008 to 2014 that second or third generation biofuels would be commercialized, most investments turned into hot air. Billions of dollars were invested in companies, including KiOR, Range Fuels, Cello Energy, Gevo, Amyris, Sun Biofuels, Solena, Solazyme, and many others went bust or have moved on to non-fuel products instead.

In the $40-$50 per barrel oil environment of 2015-2016, investments in biofuels have almost all stopped.[116] After many years of -up booms and busts, only two large-scale second generation biofuel plants are operational today, built by chemical giants with deep pockets: DuPont and Poet-DSM, both in Iowa in the United States. Both take in the leaves and stalks of corn and turn these into bioethanol, by breaking them down first into sugars with enzymes, and then fermenting these sugars using yeast. Dupont's technology has also been licensed to Chinese company Tianlong Industry, to set up a cellulosic ethanol plant in Jilin Province, China.[117]

64. How much money is needed to build a clean energy infrastructure?

The IEA and Bloomberg have evaluated global investments in new energy infrastructure. Total spending in 2015 was estimated at $1.83 trillion, or 2.5% of the world economy. The largest share of these investments flow into oil and gas extraction, refining, transport, and heating (46%), versus 17% into renewable energy, mainly electricity from hydro, solar, and wind. Other shares are spent on energy efficiency (12%), electricity networks (14%), coal and gas electricity (6%), coal extraction (4%), and nuclear power (1%).[118]–[120]

The IEA has calculated that in order to meet energy demand up to 2040, total annual investment needs to grow to $2.48 trillion per year in a 'business as usual' scenario.[121] In their scenario, growth of energy demand slows due to a doubling in energy efficiency investment, while fossil fuel use is reduced by 17%, and renewables grow nine-fold mainly for electricity.[160] The cost of an electric car transition is not yet considered by the IEA.

The impact of cost assumptions into the future for renewables and fossil fuel prices makes or breaks these forecasts. If we zoom in on power generation, both the IEA and BNEF forecast that about 60% of electricity generation capacity could be renewable by 2060, but their cost projections differ at $560 and $456 billion per year to achieve this, respectively.[118],[122],[123] Investments in clean energy are still mainly based on market support, either via government

---

160 The 'business as usual' IEA scenario calls for a total of 67 Trillion USD investment from 2014 to 2040, and the 450 ppm scenario calls for 74 trillion USD in investment.

mandates, direct subsidies, feed-in tariffs, or tax credits. Such measures are necessary to kick-start the transition and rapidly reduce technology costs; however, they are only helpful when well managed, as otherwise their costs can become too much of a burden.

Spain shows how much poorly designed regulation can affect an economy. In Spain, large parts of the electricity market, especially for poorer households, are dependent on an annual government-determined electricity price. Spain decoupled its solar and wind feed-in tariffs (FiT) in 2007 from this electricity price, removing the direct payment link.[161] In the wake of the financial crisis, the government also did not raise this electricity price, so as to not impoverish low-income households. Yet as solar power grew from 0.7 to 5.3 Gigawatts, this resulted in a giant $30 billion government deficit, an immediate end to all renewables support in 2012, and attempts to retroactively remove financial support to reduce the debt burden (and put it onto solar owners instead).[124]–[126]

As technology costs drop further, financial market support can be reduced and eventually lifted. This is already the case in some sunny and windy countries as demonstrated by the recent auctions in Morocco and the UAE for large on-shore wind and solar parks. Investments here flow because investment risk is low with a guaranteed income for 25 to 30 years.[162] The impact of removing all financial support could result in a near full stop of renewables expansion, just like we have seen in Spain and Italy.

---

161 The FiT was coupled to the Consumer Price Index instead.
162 Life-span of solar and wind infrastructure.

Epilogue

This is the second book about the world of energy I have written with Rembrandt Koppelaar. While I was responsible for the concept, title, and story line, he did most of the hard work. His research took him through enormous amounts of information, and he always came up with amazing facts, necessary for this book to come into being.

I met this brilliant young academic in 2005 in Amsterdam, when he was just eighteen years old. He contacted us, a group of concerned Dutch individuals working to set up a Dutch peak oil group. Within a year he chaired our organization.[163]

His in-depth studies of energy, starting in the early 2000s, soon made him one of the experts in this field. In an important publication about his 2006 peak oil model, he concluded:

> Given the arguments above, I do not see how liquids production peak can be postponed much further than the end of the next decade. The increase in production from the conventional resource base (including deep sea) and NGL will probably plateau/peak at the beginning of the next decade. At this point, nonconventional sources may postpone a total liquids peak for a few years, but not much longer. At this moment, I see no reason to change my peak/plateau prediction for all liquids, which stands between 2012 and 2017.[164]

163 The website is www.peakoil.nl.
164 Koppelaar (2006) Peak Oil, Separating Facts from Fiction, www.theoildrum.com/node/1889.

While the jury is still out on his prediction on 'all liquids' production, the peak of conventional oil production seems to have occurred just as he predicted. Of course not all of our predictions from our first book came true. While we rightly predicted a mass shift towards alternatives like wind and solar and the EV revolution, we didn't see the importance of the upcoming shale oil revolution. The growth of this nonconventional form of oil production, especially in the United States, has amazed even industry insiders.

That was one of the main reasons we wanted to write an update about energy markets. Both of us wanted to find out what had changed since 2008.

Yes, our backgrounds are totally different. I, as founder and fund manager of the Netherlands-based Commodity Discovery Fund, want to understand clearly the enormous impacts of changes in the world's future energy mix. It really helps us to understand the different supply and demand fundamentals for many commodities needed in this industry.

Now we understand much better why demand for copper will keep growing, platinum is needed for future fuel cell production, silver for solar panels and why oil under $50 is not sustainable. The world of energy is changing fast, and we will see many more changes shortly.

With a rising middle class of one to over two billion people, it's not hard to predict that the need for energy (and commodities) will keep growing, just as it has over the last 200 years.

Whether Tesla as a stand-alone company will survive is unsure, the revolution it helped to create can't fail. We can only hope and pray this energy transition will be completed before we lose oil as our main source of energy. Probably well before 2030, the energy to produce a barrel of oil could top the energy delivered by that same barrel. At that point we will have reached the end of our oil-based society and entered a whole new era.

I hope this book will help you to understand these important changes.

Willem Middelkoop (willem@cdfund.com)
Amsterdam, October 2016

Appendix

Table 4. World Energy Usage by Source and Use in 2015 in exajoules (EJ) and %. Values based on the author's analysis using several data sources[140],[142],[144]–[146],[148],[149],[160],[161]

All values in Exajoules	Coal	Crude Oil	Natu-ral Gas	Bio-energy	Nuclear	Hydro-power	Solar energy	Wind energy	Geo thermal
Extraction / supply	162.3	187.0	128.3	59.3	28.2	14.2	2.5	3.0	3.1
Extraction & refining	3.7	11.7	13.1	0.5	-	-	-	-	-
Inputs into Electricity	**123.8**	**13.4**	**51.7**	**5.3**	**28.2**	**14.2**	**0.9**	**3.0**	**3.0**
Power plant heat loss	90.0	9.4	32.6	1.7	18.6	-	-	-	2.5
Distribution losses	1.7	0.2	1.0	0.2	0.5	0.7	0.1	0.2	0.1
Electricity use*	31.5	4.01	19.1	3.6	8.9	14.2	0.8	2.8	0.4
Commercial & Services	7.9	0.9	4.5	0.7	2.2	3.3	0.2	1.3	0.1
Industry	14.5	1.7	8.2	1.5	4.1	6.1	0.5	0.8	0.2
Residential	9.1	1.1	5.2	1.0	2.6	3.8	0.1	0.7	0.1
Inputs into Heat	**28.9**	**17.0**	**43.5**	**48.5**	**-**	**-**	**1.5**	**-**	**0.1**
Conversion & heat losses	19.5	3.1	22.1	32.7	-	-	0.2	-	0.0
Heat use	9.4	13.9	21.4	15.8	-	-	1.3	-	0.1
Commercial & Services	1.7	3.9	1.9	0.9	-	-	0.1	-	-
Industry	3.2	0.8	5.3	3.5	-	-	0.1	-	0.1

All values in Exajoules	Coal	Crude Oil	Natu-ral Gas	Bio-energy	Nuclear	Hydro-power	Solar energy	Wind energy	Geo thermal
Residential	4.5	9.2	14.2	11.4	-	-	1.2	-	-
Transport fuels	**0.1**	**110.5**	**4.7**	**2.7**	**-**	**-**	**-**	**-**	**-**
Transport electricity	**0.6**	**0.1**	**0.4**	**0.1**	**0.2**	**0.3**	**0.0**	**0.0**	**0.0**
Fuel combustion waste heat	0.0	79.9	3.5	2.0	-	-	-	-	-
Electric vehicles waste heat	0.1	0.0	0.1	0.0	0.0	0.0	0.0	0.0	0.0
Mobility use	0.6	30.7	1.5	0.7	0.2	0.2	0.0	0.0	0.0
Light duty vehicles	0.0	13.5	1.2	0.7	0.0	0.0	0.0	0.0	0.0
Ships	-	5.0	-	-	-	-	-	-	-
Trains	0.4	0.5	0.2	0.0	0.2	0.2	0.0	0.0	0.0
Motorcycles	0.1	1.2	0.0	0.0	0.0	0.0	0.0	0.0	0.0
Heavy Trucks, Busses	0.1	5.7	0.0	0.0	0.0-	0.0	0.0	0.0	0.0
Airplanes	-	3.5	-	-	-	-	-	-	-
Other (agriculture, mining)	-	1.2	-	-	-	-	-	-	-
Non-Energy	2.2	27.4	8.2	2.3	-	-	-	-	-
Chemicals	2.2	27.4	8.2	-	-	-	-	-	-
Other	-	-	-	2.3	-	-	-	-	-
Military use / not specified	4.3	6.6	7.1	3.8	-	-	-	-	-

All values in Exajoules	Coal	Crude Oil	Natu-ral Gas	Bio-energy	Nuclear energy	Hydro-power	Solar energy	Wind energy	Geo thermal
Total final use	**48.0**	**82.3**	**56.1**	**27.6**	**9.1**	**13.5**	**2.2**	**2.8**	**0.6**
Total final use % per source	20.3%	34.8%	23.7%	9.4%	3.8%	5.7%	0.9%	1.2%	0.3%
Total electricity % per source	38.1%	4.5%	21.6%	4.1%	10.8%	16.0%	1.0%	3.4%	0.5%
Total heat % per source	15.2%	22.5%	34.6%	25.5%	-	-	2.1%	-	0.2%
Total transport % per source	1.9%	90.2%	4.4%	2.2%	0.5%	0.7%	0.0%	0.1%	0.0%

* Excluding electricity used by electric vehicles.

References Prologue

[1] Laplante, P.A. (1999). *Comprehensive Dictionary of Electrical Engineering*. Springer Science & Business Media

[2] Tesla, N. (2011). *My Inventions: The Autobiography of Nikola Tesla*. Martino Fine Books

[3] Seifer, M.J. (2016). *Wizard: The Life and Times of Nikola Tesla : Biography of a Genius*. Citadel

[4] Vujovic L., A. Marinic. (1994). *Nikola Tesla: The Genius Who Lit the World*. http://topdocumentaryfilms.com/nikola-tesla-the-genius/.

[5] Carey, C.W. (2014). *American Inventors, Entrepreneurs, and Business Visionaries*. Infobase Publishing

[6] O'Neil, J.J. (2006). *Prodigal Genius: The Life of Nikola Tesla*. Cosimo Classics

[7] Cheney, M. (2001). *Tesla: Man Out of Time*. Simon & Schuster

[8] Jonnes, J. (2004). *Empires Of Light: Edison, Tesla, Westinghouse, And The Race To Electrify The World*. Random House

[9] Carlson, W.B. (2013). *Tesla: Inventor of the Electrical Age*. Princeton University Press

[10] Tesla, N., J.T. Ratzlaff. (1984). *Tesla Said*. Tesla Book Co.

[11] Hughes, T.P. (1988). *Networks of Power: Electrification in Western Society, 1880-1930*. 2nd ed.. The John Hopkins University Press

[12] Froehlich, F.E., A. Kent, C.M. Hall. (1999). *The Froehlich/Kent Encyclopedia of Telecommunications: Volume 17*. Marcel Dekker

[13] Skrabec, Q.R. (2012). *The 100 Most Significant Events in American Business: An Encyclopedia*. Greenwood Press

[14] Skrabec, Q.R. (2006). *George Westinghouse: Gentle Genius*. Algora Publishing

[15] Naturalization index NYC Courts. (1891). *Naturalization Record of Nikola Tesla, 30 July 1891*. https://www.fold3.com/image/20564444?ann=f3dc7880-a283-11dc-2973-11792d3d4a08.

[16] Hrabak, M., R.S. Padovan, M. Kralik, D. Ozretic, K. Potocki. (2008). 'Nikola Tesla and the Discovery of X-rays'. *Radio-Graphics*. 2008;28(4):1189-1192.

[17] Tesla Universe. (2016). Tesla Timeline. Webpage. http://www. teslauniverse.com/nikola-tesla/timeline/1895-tesla-loses-fifth-avenue-lab-fire.

[18] Orton, J. (2009). *The Story of Semiconductors*. Oxford University Press

[19] Singer, P.W. (2011). *Wired for War: The Robotics Revolution and Conflict in the 21st Century*. Penguin

[20] Sarboh, S. (2006). 'Nikola Tesla's Patents'. In: *Sixth International Symposium on Nikola Tesla*. Belgrade, Serbia; 2006. https://web.archive.org/web/20071030134331/http://www. tesla-sympo6.org/papers/Tesla-Sympo6_Sarboh.pdf.

[21] Tesla, N. (1927). Tesla Patent 1,655,114 Apparatus for aerial transportation. *United States Patent Office*. http://www.teslauniverse.com/nikola-tesla/patents/us-patent-1655114-apparatus-aerial-transportation.

[22] Tesla, N. (1937). *A Machine to End War*. http://www.pbs.org/ tesla/res/res_art11.html.

[23] Castella, T. de. (2012). 'Nikola Tesla: The patron saint of geeks?' *BBC News Magazine*. http://www.bbc.co.uk/news/ magazine-19503846. Published September 10, 2012.

[24] Tesla Memorial Society of New York. (2006). The famous friends of Nikola Tesla. Webpage. http://www.teslasociety. com/famousfriends.htm.

[25] Musk, E. (2013). 'Interview with the Mind behind Tesla, SpaceX, Solarcity'. *TED*. https://www.youtube.com/ watch?v=IgKWPdJWuBQ. Published March 19, 2013.

[26] Musk, E. (2011). 'Why Invest in Making Life Multi-Planetary?' *Press Interview SpaceX*. https://www.youtube.com/ watch?v=7SECSxUbXTA. Published December 13, 2011.

[27] Andersen, R. (2014). 'Exodus: Elon Musk argues that we must put a million people on Mars if we are to ensure that humanity has a future'. *AEON*. https://aeon.co/essays/elon-musk-puts-his-case-for-a-multi-planet-civilisation. Published September 13, 2014.

[28]   Charlton, J. (2014). 'Elon Musk "Toying" with Designs for
       Electric Jet'. *Aviation.com*. http://www.aviation.com/general-
       aviation/elon-musk-toying-designs-electric-jet/. Published
       November 11, 2014.

[29]   O'Kane, S. (2015). 'Play the PC game Elon Musk
       wrote as a pre-teen'. *The Verge*. http://www.theverge.
       com/2015/6/9/8752333/elon-musk-blastar-pc-game. Pub-
       lished June 9, 2015.

[30]   Vance, A. (2016). *Elon Musk: How the Billionaire CEO of
       SpaceX and Tesla Is Shaping Our Future*. Virgin Books

[31]   Keats, R. (2013). 'Rocket Man'. *Queen's Alumni Review*. https://
       web.archive.org/web/20150504190113/http://queensu.ca/
       alumnireview/rocket-man. Published January 2013.

[32]   Friedman, J. (2003). 'Elon Musk pitches his Falcon rocket as
       a low-cost launcher, and maybe more'. *Los Angeles Times*.
       http://www.globalsecurity.org/org/news/2003/030422-
       space01.htm. Published April 22, 2003.

[33]   Compaq buys Zip2. (2002). *CNET*. https://www.cnet.com/
       news/compaq-buys-zip2/. Published January 2, 2002.

[34]   Elon Musk Biography. (2006). *Encyclopedia of World Biogra-
       phy*. http://www.notablebiographies.com/news/Li-Ou/Musk-
       Elon.html#b. Published 2006.

[35]   Musk, E. (2003). 'Success Through Viral Marketing: PayPal'.
       eCorner. http://ecorner.stanford.edu/videos/379/Success-
       Through-Viral-Marketing-PayPal. Published October 8, 2003.

[36]   O'Brien, J.M. (2007). 'The Paypal Mafia'. *Fortune Magazine*.
       http://fortune.com/2007/11/13/paypal-mafia/. Published
       November 13, 2007.

[37]   SEC U. (2002). *Form 10-K eBay Inc.* http://www.shareholder.
       com/Common/Edgar/1065088/891618-03-1538/03-00.pdf.

[38]   SEC U. (2001). *Form 10-K Paypal, INC.* https://www.sec.gov/Ar-
       chives/edgar/data/1103415/000091205702009834/a2073071z10-
       k405.htm.

[39]   Space Frontier Foundation. (2001). 'MarsNow 1.9 Profile: Elon
       Musk, Life to Mars Foundation'. *SpaceRef*. http://www.spac-
       eref.com/news/viewsr.html?pid=3698. Published September
       25, 2001.

[40] Musk, E. (2009). 'Risky Business: Why Mars is more important than cosmetics and why a failed launch is also a partial success'. *IEEE Spectrum*. http://spectrum.ieee.org/aerospace/space-flight/risky-business. Published May 30, 2009.

[41] Vance, A. (2015). 'Elon Musk's Space Dream Almost Killed Tesla'. *Bloomberg*. http://www.bloomberg.com/graphics/2015-elon-musk-spacex/. Published May 14, 2015.

[42] Bort, J. (2012). 'Here's Why Investor Steve Jurvetson Saved Elon Musk's Space Dreams'. *Business Insider*. http://www.businessinsider.com/steve-jurvetson-spacex-elon-musk-2012-9?IR=T. Published September 14, 2012.

[43] Jurvetson, S. (2014). 'SpaceX and Why They Are Daring to Think Big'. *Lecture*. https://www.youtube.com/watch?v=3aXNWGwis4w.

[44] Wayne, L. (2006). 'A Bold Plan to Go Where Men Have Gone Before'. *New York Times*. http://www.nytimes.com/2006/02/05/business/yourmoney/a-bold-plan-to-go-where-men-have-gone-before.html. Published February 5, 2006.

[45] Harwood, W. (2012). 'SpaceX Dragon returns to Earth, ends historic trip'. *CBS News*. http://www.cbsnews.com/news/spacex-dragon-returns-to-earth-ends-historic-trip/. Published May 31, 2012.

[46] Lindenmoyer, A. (2006). 'Commercial orbital transportation services (COTS) demonstrations'. *Star*. 2006;44(22). http://ezproxy.lib.ucf.edu/login?url=http://search.proquest.com/docview/23907670?accountid=10003\nhttp://sfx.fcla.edu/ucf?url_ver=Z39.88-2004&rft_val_fmt=info:ofi/fmt:kev:mtx:journal&genre=unknown&sid=ProQ:ProQ:mteabstracts&atitle=Commercial+orbital+tra.

[47] Yembrik J., J. Byerly. (2008). 'NASA Awards Space Station Commercial Resupply Services Contracts'. *NASA Contract Release*. http://www.nasa.gov/home/hqnews/2008/dec/HQ_C08-069_ISS_Resupply.html. Published December 23, 2008.

[48] Chappell, B. (2014). 'Boeing And SpaceX Win $6.8 Billion In NASA Contracts'. *NPR The Two Way*. http://www.npr.org/sections/thetwo-way/2014/09/16/349078981/boeing-and-spacex-

win-6-8-billion-in-nasa-contracts. Published September 16, 2014.

[49] 'SpaceX rocket in historic upright landing'. (2015). *BBC News*. http://www.bbc.co.uk/news/science-environment-35157782. Published December 22, 2015.

[50] O'Kane, S. (2016). 'SpaceX successfully lands a Falcon 9 rocket at sea for the third time'. *The Verge*. http://www.theverge.com/2016/5/27/11787532/spacex-falcon-9-rocket-landing-success-sea-drone-ship. Published May 27, 2016.

[51] Clark, S. (2016). 'SpaceX undecided on payload for first Falcon Heavy flight'. *Spaceflight Now*. https://spaceflightnow.com/2016/05/03/spacex-undecided-on-payload-for-first-falcon-heavy-flight/. Published May 3, 2016.

[52] Chaikin, A. (2012). 'Is SpaceX Changing the Rocket Equation?' *Air & Space Magazine*. http://www.airspacemag.com/space/is-spacex-changing-the-rocket-equation-132285884/?page=2&no-ist. Published January 2012.

[53] Clark, S. (2015). '100th Merlin 1D engine flies on Falcon 9 rocket'. *Spaceflight Now*. http://spaceflightnow.com/2015/02/22/100th-merlin-1d-engine-flies-on-falcon-9-rocket/. Published February 22, 2015.

[54] Cofield, C. 'SpaceX Will Launch Private Mars Missions as Soon as 2018'. *Space.com*. http://www.space.com/32719-spacex-red-dragon-mars-missions-2018.html. Published April 27, 2016.

[55] Davenport, C. (2016). 'Elon Musk provides new details on his "mind blowing" mission to Mars'. https://www.washingtonpost.com/news/the-switch/wp/2016/06/10/Elon-musk-provides-new-details-on-his-mind-blowing-mission-to-mars/. Published June 10, 2014.

[56] Baer, D. (2014). 'The Making of Tesla: Invention, Betrayal, and the Birth of the Roadster'. *Business Insider*. http://uk.businessinsider.com/tesla-the-origin-story-2014-10?r=US&IR=T. Published November 11, 2014.

[57] Morrison, C. (2008). 'Elon Musk steps in as CEO at Tesla, lays off staff'. *New York Times*. http://www.nytimes.com/external/venturebeat/2008/10/15/15venturebeat-elon-musk-steps-in-

as-ceo-at-tesla-lays-off-99182.html. Published October 15, 2008.

[58] Musk, E. (2014). 'All our Patent Are Belong to You'. *Tesla Webpage*. http://www.teslamotors.com/blog/all-our-patent-are-belong-you. Published June 12, 2014.

[59] Solarcity. (2016). Solarcity Executive Management. Webpage. http://www.solarcity.com/company/team.

[60] Gornall, J. (2016). 'Newsmaker: Elon Musk'. *The National*. http://www.thenational.ae/arts-life/newsmaker-elon-musk. Published August 4, 2016.

[61] Solar Power World. (2016). 2016 Top 500 North American Solar Contractors. Webpage. http://www.solarpowerworldonline.com/2016-top-500-north-american-solar-contractors/.

[62] Diggelen, A. van. (2012). 'Tesla and SolarCity Collaborate on Clean Energy Storage'. *KQED*. http://blogs.kqed.org/climate-watch/2012/04/24/tesla-and-solarcity-collaborate-on-clean-energy-storage/. Published April 24, 2012.

[63] Tesla Motors. (2006). *The Unveiling of the Tesla Roadster*. https://www.youtube.com/watch?v=hOl_1S10jTk.

[64] Vance, A. (2013). 'Revealed: Elon Musk Explains the Hyperloop, the Solar-Powered High-Speed Future of Inter-City Transportation'. *Bloomberg*. http://www.bloomberg.com/news/articles/2013-08-12/revealed-elon-musk-explains-the-hyperloop-the-solar-powered-high-speed-future-of-inter-city-transportation. Published August 12, 2013.

References Chapter 1

[1] Bank, T.W. (2016). Infrastructure reliability and availability dataset. World Bank Group Enterprise Surveys. http://www.enterprisesurveys.org/data/exploretopics/infrastructure.

[2] United Nations Population Division. (2016). World Population Prospects, the 2015 Revision. https://esa.un.org/unpd/wpp/Download/Standard/Population/.

[3] Heerman, K. (2016). Real GDP (2010 dollars) Historical Dataset. US Department of Agriculture Economic Research

Service. http://www.ers.usda.gov/data-products/international-al-macroeconomic-data-set.aspx.

[4] International Energy Agency (IEA). (2016). *Energy and Air Pollution*.

[5] Putin, V. (2015). Speech Vladimir Putin at the Paris Climate Conference. In: *UNFCCC*. http://unfccc.int/meetings/paris_nov_2015/items/9331.php.

[6] Harvey F, Taylor L. (2015). Russia pledges not to stand in the way of Paris Climate deal. *The Guardian*. https://www.theguardian.com/environment/2015/dec/07/russia-pledges-not-to-stand-in-the-way-of-paris-climate-deal. Published December 7, 2015.

[7] Liebreich, M. (2016). *Bloomberg New Energy Finance Summit*.

[8] IEA. (2016). *World Energy Investment 2016 Fact Sheet*.

[9] Trip, R. (2009). Interview Jeroen van der Veer. *Buitenhof*. http://www.npo.nl/buitenhof/22-02-2009/VPRO_1131491.

[10] Pashley, A. (2015). Solar power "backbone" of future energy system – Shell CEO. Climate Home. http://www.climatechangenews.com/2015/09/17/solar-power-backbone-of-future-energy-system-shell-ceo/.

[11] Lambert, F. (2016). 'Tesla Model 3: There's a way to see where you are in the queue, check it out before Tesla finds out'. *Electrek*. https://electrek.co/2016/06/07/tesla-model-3-reservation-queue-number/. Published June 7, 2016.

[12] Youtube. (2011). Elon Musk Thoughts on Transitioning to 100% renewable Energy. https://www.youtube.com/watch?v=HiOLan8JocE.

[13] Smil, V. (2005). *Creating the Twentieth Century: Technical Innovations of 1867-1914 and Their Lasting Impact* (1st ed.). Oxford University Press

[14] Smil, V. (2007). 'Prime movers of globalization: The history and impact of diesel engines and gas turbines'. *J Glob Hist*. 2007;(2):373-394. DOI:10.1017/S1740022807002331.

[15] Carstens, H. (2008). 'The Discovery that Changed the Oil Industry Forever'. *Geo ExPro*. 2008;(May). http://assets.geo-expro.com/legacy-files/articles/History of Oil Discovery That Changed Oil For Ever Spindletop.pdf.

[16]   Gordon, R.J. (2016). *The Rise and Fall of American Growth* (1st ed.). Princeton University Press

[17]   Lutz, B. (2013). *Car Guys vs. Bean Counters: The Battle for the Soul of American Business*. Portfolio

[18]   Zee, B. van der. (2010). 'Tesla's Roadster Sport Saves the Electric Car'. *The Guardian*. https://www.theguardian.com/environment/green-living-blog/2010/feb/03/tesla-roadster-sport-electric-car. Published February 3, 2010.

[19]   Musk, E. (2011). 'Why Invest in Making Life Multi-Planetary?'. *Press Interview SpaceX*. https://www.youtube.com/watch?v=7SECSxUbXTA. Published December 13, 2011.

[20]   Musk, E. (2013). 'Interview with the Mind behind Tesla, SpaceX, Solarcity'. *TED*. https://www.youtube.com/watch?v=IgKWPdJWuBQ. Published March 19, 2013.

[21]   Andersen, R. (2014). 'Exodus: Elon Musk argues that we must put a million people on Mars if we are to ensure that humanity has a future'. *AEON*. https://aeon.co/essays/elon-musk-puts-his-case-for-a-multi-planet-civilisation. Published September 13, 2014.

[22]   Baer, D. (2014). 'The Making Of Tesla: Invention, Betrayal, and The Birth of the Roadster'. *Business Insider*. http://uk.businessinsider.com/tesla-the-origin-story-2014-10?r=US&IR=T. Published November 11, 2014.

[23]   Morrison, C. (2008). 'Elon Musk steps in as CEO at Tesla, lays off staff'. *Venture Beat*. http://venturebeat.com/2008/10/15/elon-musk-steps-in-as-ceo-at-tesla-lays-off-staff/. Published October 15, 2008.

[24]   Vance, A. (2016). *Elon Musk: How the Billionaire CEO of SpaceX and Tesla Is Shaping Our Future*. Virgin Books

[25]   Assis, C. (2016). Elon Musk exercises Tesla options, pays $50 million tax bill with own cash. *Market Watch*. http://www.marketwatch.com/story/elon-musk-buys-tesla-shares-cheap-pays-hefty-tax-bill-with-own-cash-2016-01-29. Published January 30, 2016.

[26]   Ziegler C. (2016). 'Elon Musk bought $100 million more worth of Tesla this week'. *The Verge*. http://www.theverge.com/2016/1/29/10873576/elon-musk-100-million-option-exercise-stock-tesla. Published January 29, 2016.

[27] Tesla Motors. (2016). Model 3 Accelerating Sustainable Transport. Tesla Motors Webpage. https://www.tesla.com/en_GB/model3?redirect=no. Published 2016.

[28] Schonfeld, E. (2009). 'The Government Comes Through for Tesla with a $465 Million Loan for Its Electric Sedan'. *Tech Crunch*. https://techcrunch.com/2009/06/23/the-government-comes-through-for-tesla-with-a-465-million-loan-for-its-electric-sedan/. Published June 23, 2009.

[29] Hull, D, T. Black (2016). 'Tesla Jumps as Third-Quarter Shipments Aid Musk Funding Plan'. *Bloomberg*. https://www.bloomberg.com/news/articles/2016-10-02/tesla-delivered-24-500-vehicles-in-third-quarter-cnbc-reports. Published October 2, 2016.

[30] OICA. (2015). OICA World Motor Vehicle Sales. http://www.oica.net/category/sales-statistics/.

[31] Rechtin, M. (2015). 'Tesla Model S P85D Earns Top Road Test Score'. *Consumer Reports*. http://www.consumerreports.org/cro/cars/tesla-model-s-p85d-earns-top-road-test-score. Published October 20, 2015.

[32] Consumer Reports. (2016). 'Tesla Model S'. *Consumer Reports Website*. http://www.consumerreports.org/cro/tesla-model-s.htm.

[33] Automotive Science Group. (2016). 2016 Automotive Performance Index Findings. *Automotive Science Group Webpage*. http://www.automotivescience.com/pages/the-study.

[34] Loveday, E. (2014). Nissan LEAF Has Smallest Lifecycle Footprint of Any 2014 Model Year Automobile Sold in North America. *InsideEVs*. http://insideevs.com/nissan-leaf-has-smallest-lifecycle-footprint-of-any-2014-automobile-sold-today-in-north-america/. Published February 11, 2014.

[35] Loveday, E. (2016). 'Tesla Ups Supercharger Charging Rate For Refreshed Model S 90D & P90D – Video'. *InsideEvs*. http://insideevs.com/tesla-ups-supercharger-charging-rate-refreshed-model-s-90d-p90d-video/. Published July 10, 2016.

[36] Tesla Motors. (2016). 'Supercharger'. *Tesla Motors Webpage*. https://www.tesla.com/en_GB/supercharger?redirect=no.

[37] Fowler, S. (2014). 'Elon Musk: Tesla boss on EVs with 500-mile range and colonies on Mars'. *Auto Express*. http://www.

autoexpress.co.uk/tesla/87943/elon-musk-tesla-boss-on-evs-with-500-mile-range-and-colonies-on-mars. Published July 21, 2014.

[38] Weberm, H. (2016). 'Elon Musk: Tesla Model 3 owners won't get free Supercharging for life without paying extra'. *Venture Beat*. http://venturebeat.com/2016/05/31/elon-musk-model-3-owners-wont-get-free-supercharging-for-life-by-default/. Published May 31, 2016.

[39] Hanley, S. (2016). 'Tesla Planning More Superchargers Across Europe in 2016'. *TeslaRati*. http://www.teslarati.com/tesla-planning-superchargers-europe-2016/. Published January 16, 2016.

[40] ENS Economic Bureau. (2016). 'Tesla Model 3 Coming to India; Musk Says Will Throw in India-wide Supercharger Network'. *The New Indian Express*. http://www.newindianexpress.com/cities/chennai/2016/apr/01/Tesla-Model-3-Coming-to-India-Musk-Says-Will-Throw-in-India-wide-Supercharger-Network-918852.html. Published April 1, 2016.

[41] Motors T. (2015). Tesla Gigafactory. https://www.teslamotors.com/en_GB/gigafactory.

[42] Wesoff, E. (2016). 'How Soon Can Tesla get Battery Cell Costs Below 100 USD per Kilowatt-Hour'. *Greentech Media*. http://www.greentechmedia.com/articles/read/How-Soon-Can-Tesla-Get-Battery-Cell-Cost-Below-100-per-Kilowatt-Hour.

[43] The Tesla Team. (2016). 'Tesla Makes Offer to Acquire Solar-City'. *Tesla Motors Webpage*. https://www.tesla.com/en_GB/blog/tesla-makes-offer-to-acquire-solarcity?redirect=no. Published June 21, 2016.

[44] Musk, E. (2014). 'All our Patent Are Belong to You'. *Tesla Motors Webpage*. http://www.teslamotors.com/blog/all-our-patent-are-belong-you. Published June 12, 2014.

[45] Gordon-Bloomfield, N. (2016). 'As Global Nissan LEAF Odometer Passes 1 Billion Miles, Nissan Celebrates Sale of 10,000th LEAF in UK'. *Transport Evolved*. https://transportevolved.com/2015/07/06/as-global-nissan-leaf-odometer-passes-1-billion-miles-nissan-celebrates-sale-of-10000th-leaf-in-uk/. Published July 5, 2016.

[46] Cole, J. (2015). Tesla Model S Fleet Passes 1 Billion Miles Driven – Video. *InsideEVs*. http://insideevs.com/tesla-model-s-passes-1-billion-miles-driven/. Published June 23, 2015.

[47] Loveday, E. (2014). 'Chevrolet Celebrates Volt's 1 Billion Mile Milestone'. *InsideEVs*. http://insideevs.com/chevrolet-celebrates-volts-1-billion-mile-milestone/. Published October 5, 2014.

[48] Epstein, Z. (2016). 'Elon Musk says Tesla's next car will be even cheaper than the Model 3'. *BGR*. http://bgr.com/2016/04/26/tesla-model-4-price-elon-musk/. Published April 26, 2016.

[49] BSW-Solar. (2016). Preisindex Photovoltaik des 1. Quartals 2016. Preisindex Photovoltaik. https://www.solarwirtschaft.de/preisindex.html.

[50] Thalman, E. (2016). 'What German households pay for power'. *Clean Energy Wire*. https://www.cleanenergywire.org/factsheets/what-german-households-pay-power. Published July 26, 2016.

[51] Pyper, J. (2016). 'Sonnen Launches a Home Battery for Self-Consumption at a 40% Reduced Cost'. *Green Tech Media*. http://www.greentechmedia.com/articles/read/Sonnen-Launches-a-Home-Battery-for-Self-Consumption-at-a-40-Reduced. Published July 7, 2016.

[52] Sonnen. (2016). Sonnenbatterie. Sonnen. https://www.sonnen-batterie.com/en-us/start.

[53] Borgmann, M. (2014). 'Dubai's DEWA procures the world's cheapest solar energy ever: Riyadh, start your photocopiers'. *Apricum Group*. November 27, 2014.

[54] Graves, L. (2015). 'UAE beats renewables cost hurdle with world's cheapest price for solar energy'. *The National*. January 18, 2015.

[55] Borgmann, M. (2016). 'Dubai Shatters all Records for Cost of Solar with Earth's Largest Solar Power Plant'. *Apricum*. http://www.apricum-group.com/dubai-shatters-records-cost-solar-earths-largest-solar-power-plant/. Published May 2, 2016.

[56] US Department of Energy. (2014). 'U.S. Utility-Scale Solar 60 Percent Towards Cost-Competition Goal'. *News Media Energy.Gov*. http://energy.gov/articles/us-utility-scale-solar-

60-percent-towards-cost-competition-goal. Published February 12, 2014.

[57] Lacey, S. (2015). 'Cheapest Solar Ever: Austin Energy Gets 1.2 Gigawatts of Solar Bids for Less than 4 Cents'. *Greentech Media*. https://www.greentechmedia.com/articles/read/cheapest-solar-ever-austin-energy-gets-1.2-gigawatts-of-solar-bids-for-less. Published June 30, 2015.

[58] PTI. (2016). 'Solar power tariff at record low, drops to Rs 4.34 a unit'. *The Economic Times*. http://articles.economictimes.indiatimes.com/2016-01-19/news/69900231_1_power-tariff-sunedison-ntpc. Published January 19, 2016.

[59] Bloomberg Energy Intelligence. (2016). BNEF Multicrystalline Module Price.

[60] International Technology Roadmap for Photovoltaic Association. (2016). *International Technology Roadmap for Photovoltaic (ITRPV) 2015 Results*. Vol Seventh ed. papers2://publication/uuid/20F56C7C-3684-4039-B043-D3DE7C5293FA.

[61] Siemer J, Knoll B. (2013). Still More Than Enough. *Phot Int*. February 2013:73.

[62] Bloomberg Energy Intelligence. (2016). BNEF Polysilicon price. 2016.

[63] Chaintore, P.V., I. Gordon, W. Hoffmann et al. (2015). *Future Renewable Energy Costs: Solar Photovoltaics*.

[64] Polman, A., M. Knight, E.C. Garnett, B. Ehrler, W.C. Sinke. *Photovoltaic materials – present efficiencies and future challenges*:1-24. DOI:10.1126/science.aad4424.

[65] LaMonica, M. (2011). 'Silevo's hybrid solar cell challenges status quo'. *CNET*. http://www.cnet.com/uk/news/silevos-hybrid-solar-cell-challenges-status-quo/. Published October 12, 2011.

[66] Martin, R. (2016). 'Solarcity's Gigafactory: a 750 million USd solar facility in Buffalo will produce a gigawatt of high efficiency solar panels per year and make the technology far more attractive to homeowners'. *MIT Technology Review*. https://www.technologyreview.com/s/600770/10-breakthrough-technologies-2016-solarcitys-gigafactory/.

[67] Cooper, K. (2015). 'Solarcity unveils world's most efficient rooftop solar panel, to be made in America'. *Solarcity*. http://

www.solarcity.com/newsroom/press/solarcity-unveils-world%E2%80%99s-most-efficient-rooftop-solar-panel-be-made-america. Published October 2, 2015.

[68] Martin, C. (2016). 'Elon Musk Is Trying to Perfect Solar Roof Tiles'. *Bloomberg*. http://www.bloomberg.com/news/articles/2016-10-28/musk-beguiled-by-solar-roofing-faces-battle-many-have-lost. Published October 28, 2016.

[69] IEA. (2016). *Global EV Outlook 2016 Beyond One Million Electric Cars*.

[70] Nationale Plattform Elektromobilität. (2012). *Fortschrittsbericht Der Nationalen Plattform Elektromobilität (Dritter Bericht)*. http://www.bmwi.de/BMWi/Redaktion/PDF/Publikationen/fortschrittsbericht-der-nationalen-plattform-elektromobilitaet.

[71] Bundesministerium fur Umwelt, Naturschutz B und R. (2014). *Erneuerbar mobil:Marktfähige Lösungen Für Eine Klimafreundliche Elektromobilität*.

[72] The Editors. (2016). 'Kaufprämien und Steuerboni für Elektroautos'. *Taguesschau.de*. http://www.tagesschau.de/wirtschaft/elektro-autos-103.html. Published May 15, 2016.

[73] Bloomberg News. (2016). 'China Plans to End New Energy Vehicles Subsidies after 2020'. *Bloomberg*. http://www.bloomberg.com/news/articles/2016-01-23/china-plans-to-end-new-energy-vehicles-subsidies-after-2020. Published January 23, 2016.

[74] PTI. (2016). 'Plans afoot to make India 100% e-vehicle nation by 2030: Piyush Goyal'. *The Economic Times of India*. http://articles.economictimes.indiatimes.com/2016-03-26/news/71829886_1_piyush-goyal-oil-minister-dharmendra-pradhan-distribution. Published March 26, 2016.

[75] J.C. (2016). 'Norway Aiming For 100-Percent Zero Emission Vehicle Sales By 2025'. *hybridCars*. http://www.hybridcars.com/norway-aiming-for-100-percent-zero-emission-vehicle-sales-by-2025/. Published March 8, 2016.

[76] Electric Vehicles Initiative. (2013). *Global EV Outlook: Understanding the Electric Vehicle Landscape to 2020*.

[77] Jose, P. (2016). EV Sales – World Top 20 December 2015 Special Edition. EV Sales. http://ev-sales.blogspot.co.uk/2016_01_01_archive.html.

[78] Randall, T. (2016). 'Here's how electric cars will cause the next oil crisis'. *Bloomberg*. http://www.bloomberg.com/features/2016-ev-oil-crisis/.

[79] Lienert P. (2015). 'Automakers race to double the driving range of affordable electric cars'. *Reuters*. http://www.reuters.com/article/us-autos-batteries-idUSKBN0MK2FC20150324. Published March 4, 2015.

[80] Cobb, J. (2016). 'Five Pending 200-Mile Range EVs that Won't Break the Bank'. *hybridCars*. http://www.hybridcars.com/five-pending-200-mile-range-evs-that-wont-break-the-bank/.

[81] Daisuke, W. (2015). 'Apple Targets Electric Car Shipping Date for 2019'. *Wall Street Journal*. http://www.wsj.com/articles/apple-speeds-up-electric-car-work-1442857105. Published September 21, 2015.

[82] Muoio, D. (2016). 'Henrik Fisker is using a revolutionary new battery to power his Tesla killer'. *Business Insider UK*. http://uk.businessinsider.com/henrik-fisker-is-using-revolutionary-battery-tech-for-electric-car-2016-10. Published October 17, 2016.

[83] Carrington, D. (2016). 'Dyson could become next Tesla with its electric car, says expert'. *The Guardian*. https://www.theguardian.com/environment/2016/may/11/dysons-electric-car-development-could-become-the-next-tesla. Published May 11, 2016.

[84] Bloomberg New Energy Finance. (2016). Electric Vehicles to be 35% of Global New Car Sales By 2040. http://about.bnef.com/press-releases/electric-vehicles-to-be-35-of-global-new-car-sales-by-2040/.

[85] Statista. (2016). Statista – The Statistics Portal. http://www.statista.com/.

[86] Pontes, J. (2016). EV Sales. EV Sales. http://ev-sales.blogspot.co.uk/.

[87] Hanley, S. (2015). 'Nissan predicts 10% electric car sales by 2020'. *Ecomento*. http://ecomento.com/2015/10/22/nissan-

predicts-10-electric-car-sales-by-2020/. Published October 2015.

[88] Cobb, J. (2015). 'Nissan's Breakthrough Battery Will Enable 10-Percent EV sales by 2020'. *hybridCARS*. October 2015. http://www.hybridcars.com/nissans-breakthrough-battery-will-enable-10-percent-ev-sales-by-2020/.

[89] Daniela, W. (2016). 'BYD Sees electric-car sales tripling in market coveted by Tesla'. *Bloomberg News*. http://www.bloomberg.com/news/articles/2016-03-29/byd-sees-electric-car-sales-tripling-in-market-coveted-by-tesla. Published March 29, 2016.

[90] Kane, M. (2016). 'BYD Increases Net Earnings and Expects To Double Plug-In Sales Every Year Through 2018'. *Inside EVs*. http://insideevs.com/byd-increases-net-earnings-and-expects-to-double-plug-in-sales-every-year-through-2018/.

[91] AFP. (2016). 'Tesla accelerates to hit target of making 500,000 cars'. *Channel NewsAsia*. http://www.channelnewsasia.com/news/business/tesla-accelerates-to-hit/2757308.html. Published May 5, 2016.

[92] Lambert, F. (2016). 'VW announces new plan to build 2 to 3 million all-electric cars a year by 2025'. *electrek*. June 2016. http://electrek.co/2016/06/16/vw-2-3-million-all-electric-cars-2025/.

[93] Herndon, V. (2015). 'Audi declares at least 25% of US sales will come from electric vehicles by 2025'. *Audi Press Release*. November 18, 2015.

[94] Joseph, N. (2015). 'Geely Emgrand EV kicks off transition to electric propulsion'. *Autoblog*. November 2015. http://www.autoblog.com/2015/11/18/geely-emgrand-ev-official/.

[95] Benny. (2015). 'BMW i5: Kruger confirms new electric car above the i3'. *Bimmer Today*. http://www.bimmertoday.de/2015/10/22/bmw-i5-kruger-bestatigt-neues-elektroauto-oberhalb-des-i3/. Published October 2015.

[96] Ford. (2015). Ford Investing 4.5 Billion USD in Electrified Vehicle Solutions Reimagining How to Create Future Vehicle User Experiences. Ford Motor Company Media Center. https://media.ford.com/content/fordmedia/fna/us/en/

news/2015/12/10/ford-investing-4-5-billion-in-electrified-vehicle-solutions.html.

[97] Ben, K., Z. Nick. (2015). 'GM expects to fall short of 2017 electric vehicles targets'. *Reuters*. May 8, 2015.

[98] Jie, M. (2016). 'BAIC Predicts 35-Fold Surge in Electric Car Sales on State Push'. *Bloomberg News*. http://www.bloomberg.com/news/articles/2016-01-13/baic-predicts-35-fold-surge-in-electric-car-sales-on-state-push. Published January 13, 2016.

[99] Korosec, K. (2015). 'Volvo: It's time for electric cars to enter the mainstream. *Fortune Mag*. October 2015. http://fortune.com/2015/10/15/volvo-electric-cars-lineup/.

[100] Schaal, E. (2016). 'Hyundai Electric Vehicle will Raise Low-End Bar to 250 Miles'. *Autos CheatSheet*. http://www.cheatsheet.com/automobiles/hyundai-electric-vehicle-250-miles.html/?a=viewall. Published June 2016.

[101] Hachigo, T. (2016). 'Honda CEO Shifts Focus to Electric Vehicles'. *The Wall Street Journal*. Published February 25, 2016.

[102] Edelstein, S. (2013). 'How does GM's fabled EV1 stack up against the current crop of electrics?'. *Digital Trends*. Published February 28, 2013.

[103] Sedgwick, D. (1996). 'Saturn and EV1 Buyers Make Odd Coupling'. *Automotive News*. http://www.autonews.com/article/19960205/ANA/602050771/saturn-and-ev1-buyers-make-odd-coupling. Published February 5, 1996.

[104] Nykvist, B., M. Nilsson. (2015). 'Rapidly falling costs of battery packs for electric vehicles'. *Nat Clim Chang*. 2015;5(4):329-332. DOI:10.1038/nclimate2564.

[105] Yuzawa, K., Y. Yang, N. Bhandari, M. Sugiyama, S. Nakamura. (2016). 'Charging the future: Asia leads drive to next-generation EV battery market'. *Goldman Sachs Equity Research*. Published September 27, 2016.

[106] Lambert, F. (2016). 'Tesla confirms base Model 3 will have less than 60 kWh battery pack option, cost is below $190/kWh and falling'. *Electrek*. https://electrek.co/2016/04/26/tesla-model-3-battery-pack-cost-kwh/. Published April 4, 2016.

[107] Zu, C-X., H. Li. (2011). 'Thermodynamic analysis on energy densities of batteries'. *Energy Environ Sci*. 2011;4(8):2614. DOI:10.1039/c0ee00777c.

[108] Bruce, P.G., S.A. Freunberger, L.J. Hardwick, J-M. Tarascon. (2011). 'Li–O2 and Li–S batteries with high energy storage'. *Nat Mater*. 2011;11(02):172-172. DOI:10.1038/nmat3237.

[109] Thackeray, M.M., C. Wolverton, E.D. Isaacs. (2012). 'Electrical energy storage for transportation—approaching the limits of, and going beyond, lithium-ion batteries'. *Energy Environ Sci*. 2012;5(7):7854. DOI:10.1039/c2ee21892e.

[110] White, J. (2016). 'Tesla's Musk says master plan will require capital raise'. *Reuters*. http://uk.reuters.com/article/uk-tesla-gigafactory-idUKKCN1062T1. Published July 27, 2016.

[111] Cobb, J. (2015). 'Tesla Projects Battery Costs Could Drop To $100/KWH By 2020'. *hybridCars*. http://www.hybridcars.com/tesla-projects-battery-costs-could-drop-to-100kwh-by-2020/. Published June 18, 2015.

[112] Buchmann, I. (2015). 'Battery Statistics'. *Batter Univ*. 2015:1-12. http://batteryuniversity.com/learn/article/battery_statistics.

[113] Yonhap News Agency. (2015). 'Hyundai Tucson Fuel Cell Sales Not Hitting Target'. *Autoblog*. http://www.autoblog.com/2015/06/17/hyundai-tucson-fuel-cell-sales-missing-target/. Published June 17, 2015.

[114] Shelton, S. (2015). 'Toyota Mirai goes on sale with 2000 preorders'. *hybridCars*. http://www.hybridcars.com/toyota-mirai-goes-on-sale-with-2000-preorders/. Published October 26, 2015.

[115] Bloomberg. (2015). 'Honda Takes on Toyota Mirai with roomier 63,000 USD fuel cell sedan'. *Bloomberg*. http://www.bloomberg.com/news/articles/2015-10-28/honda-takes-on-toyota-mirai-with-roomier-63-000-fuel-cell-sedan. Published October 28, 2015.

[116] Krok, A. (2016). 'Mercedes' GLC F-Cell is the first plug-in fuel cell vehicle ever'. *Road Show*. http://www.cnet.com/road-show/news/mercedes-glc-f-cell-is-the-first-plug-in-fuel-cell-vehicle-ever/. Published June 13, 2016.

[117] Autoblog. (2016). 'Mercedes-Benz GLC plug-in hydrogen fuel-cell coming in 2017'. *Autoblog*. http://www.autoblog.

com/2016/06/13/mercedes-benz-glc-plug-in-hydrogen-fuel-cell-coming-in-2017/. Published June 13, 2016.

[118] Tajitsu, N. (2015). 'BMW plans to market sedan fuel cell vehicle after 2020'. *Reuters*. http://www.reuters.com/article/autoshow-japan-bmw-idUSL3N12T3TN20151029. Published October 29, 2015.

[119] Healey, T. (2016). 'Honda and GM share 2020 vision on jointly developed fuel cell vehicles'. *hybridCars*. http://www.hybridcars.com/honda-and-gm-share-2020-vision-on-jointly-developed-fuel-cell-vehicles/. Published March 14, 2016.

[120] Stelton, S. (2015). 'Nissan to Launch Fuel Cell Vehicle by 2021'. *hybridCars*. http://www.hybridcars.com/nissan-to-launch-fuel-cell-vehicle-by-2021/. Published November 6, 2015.

[121] Tajitsu, N. (2015). 'Toyota targets fuel-cell car sales of 30,000 a year by 2020'. *Reuters*. http://www.reuters.com/article/us-toyota-environment-idUSKCN0S80B720151014. Published October 14, 2015.

[122] Quiroga, T. (2014). '2015 Hyundai Tucson Fuel Cell'. *Car and Driver*. http://www.caranddriver.com/reviews/2015-hyundai-tucson-fuel-cell-first-drive-review. Published November 2014.

[123] Bossel, U. (2006). 'Does a hydrogen economy make sense?'. *Proc IEEE*. 2006;94(10):1826-1836. DOI:10.1109/JPROC.2006.883715.

[124] Hwang, J-J. (2013). 'Sustainability study of hydrogen pathways for fuel cell vehicle applications'. *Renew Sustain Energy Rev*. 2013;19:220-229. DOI:10.1016/j.rser.2012.11.033.

[125] NPR. (2008). Al Gore's speech on renewable energy. http://www.npr.org/templates/story/story.php?storyId=92638501. Published July 17, 2008.

[126] Koppelaar, R. (2012). 'World Energy Consumption – Beyond 500 Exajoules'. *The Oil Drum*. http://www.theoildrum.com/node/8936. Published February 16, 2012.

[127] Smil, V. (2014). 'The long slow rise of solar and wind: the great hope for a quick and sweeping transition to renewable energy is wishful thinking'. *Sci Am*. 2014:52-57. http://www.vaclavsmil.com/wp-content/uploads/scientificameri-can0114-521.pdf.

[128] Smil, V. (2008). 'Moore's Curse and the Great Energy Delusion'. *Am*. November 2008. http://www.vaclavsmil.com/wp-content/uploads/docs/smil-article-20081119-the_American.pdf.

[129] Bloomberg. (2016). 'Goldwind eyes mass production of 6MW turbines in three years'. *Energy & Commodities*. http://www.businesstimes.com.sg/energy-commodities/goldwind-eyes-mass-production-of-6mw-turbines-in-three-years. Published August 11, 2016.

[130] Dai, Y. (2014). *The Innovation Path of the Chinese Wind Power Industry*.

[131] Vensys. (2015). Vensys Then and Now. http://www.vensys.de/energy-en/unternehmen/historie.php. Published 2015.

[132] Clark, P. (2016). 'China's Goldwind becomes world's largest wind turbine maker'. *Financial Times*. https://www.ft.com/content/123f1af0-d97e-11e5-a72f-1e7744c66818. Published February 23, 2016.

[133] Musk, E. (2014). 'All Our Patent Are Belong to You'. *Press Release Tesla Motors*. https://www.tesla.com/en_GB/blog/all-our-patent-are-belong-you?redirect=no. Published June 12, 2014.

[134] Toyota. (2015). 'Toyota opens the door and invites the industry to the hydrogen future'. *Press Release*. http://corporate-news.pressroom.toyota.com/releases/toyota+fuel+cell+patents+ces+2015.htm. Published January 5, 2015.

[135] Nichols, W. (2015). 'Ford opens up electric vehicle patents to rivals'. *businessGreen*. http://www.businessgreen.com/bg/news/2410656/ford-opens-up-electric-vehicle-patents-to-rivals. Published May 29, 2015.

[136] Masnick, M. (2015). 'Elon Musk Clarifies that Tesla's Patents Really Are Free'. *tech dirt*. https://www.techdirt.com/articles/20150217/06182930052/elon-musk-clarifies-that-teslas-patents-really-are-free-investor-absolutely-freaks-out.shtml. Published February 18, 2015.

[137] IEA, Nuclear Energy Agency. (2015). *Projected Costs of Generating Electricity*. DOI:10.1787/9789264084315-en.

[138] Trend:Research, Leuphana Universität Lüneburg. (2013). Definition und Marktanalyse von Bürgerenergie in Deutschland. 2013:76.

[139] Perlin, J. (2016). *Let It Shine The 6000 Year Story of Solar Energy*. (2nd ed.). New World Library

[140] IEA. (2016). *Key Electricity Trends Excerpt from Electricity Information 2016*.

[141] IEA. (2016). *Key Coal Trends Excerpt from Coal Information 2016*.

[142] IEA. (2016). *Key Natural Gas Trends Excerpt from Natural Gas Information 2016*.

[143] Zervos, A., C. Lins. (2016). *Renewables 2016: Global Status Report*.

[144] IEA. (2015). *Energy Balances of Non-OECD Countries*.

[145] IEA. (2015). *Energy Balances of OECD Countries*.

[146] IEA. (2016). *Key Renewables Trends Excerpt from: Renewables Information 2016*.

[147] IEA. (2015). *Key Renewables Trends Excerpt from: Renewables Information 2015*.

[148] International Renewable Energy Agency. (2016). *Renewable Energy Statistics 2016*.

[149] BP. (2016). BP Statistical Review of World Energy June 2016. http://www.bp.com/statisticalreview.

[150] Heerman, K. (2016). ERS International Macroeconomic Data Set. USDA. http://www.ers.usda.gov/data-products/international-macroeconomic-data-set.aspx. Published 2016.

[151] ExxonMobil. (2016). *Energy: A View to 2040*.

[152] British Petroleum. (2016). *BP Energy Outlook 2016 Edition*.

[153] IEA. (2015). *World Energy Outlook 2015*.

[154] Statista. (2016). Total length of public roads in China from 2005 to 2015 (in million kilometers). *Statista*. http://www.statista.com/statistics/276051/total-length-of-public-roads-in-china/. Published May 20, 2016.

[155] Statista. (2016). Length of China Railway's rail network from 2008 to 2015, by railroad embankment type (in 1,000 kilometers). *Statista*. May 2016. http://www.statista.com.iclibezp1.cc.ic.ac.uk/statistics/276254/length-of-china-railways-rail-network-by-railroad-embankment-type/.

[156] Bekker, G.A., R.T. Segers, H. Zhang. (2007). *Energy in China: An Introduction to China and Its Contemporary Energy Situation*. (1st ed.). Energy Delta Institute.

[157] China National Energy Board. (2016). National Energy Board total Electricity Consumption Figures March 2016. Electricity Consumption Figures. http://www.nea.gov.cn/2016-04/15/c_135282340.htm. Published 2016.

[158] Bloomberg. (2016). China Steel Production from World Bureau of Metal Statistics. *Bloom Database*. March 2016.

[159] Bloomberg. (2016). China Cement Production from Chinese Government Data. *Bloom Database*. April 2016.

[160] IEA. (2015). WEO 2015 Electricity Access Database. http://www.worldenergyoutlook.org/resources/energydevelopment/energyaccessdatabase/. Published 2015.

[161] IEA. (2015). *Coal Information 2015*.

[162] Speirs, J., M. Contestabile, Y. Houari, R. Gross. (2014). 'The future of lithium availability for electric vehicle batteries'. *Renew Sustain Energy Rev*. 2014;35(August 2016):183-193. DOI:10.1016/j.rser.2014.04.018.

[163] Gordon, R. (2015). 'Cobalt and the Tesla effect: higher prices or a risk of substitution?'. *CRU Group*. http://www.crugroup.com/about-cru/cruinsight/Cobalt_and_the_Tesla_effect_Higher_Prices_or_a_risk_of_substitution. Published May 6, 2015.

[164] Desai, P. (2016). 'RPT-Electric vehicles to power cobalt revival'. *Reuters*. http://www.reuters.com/article/metals-cobalt-demand-idUSL8N1903ML. Published June 9, 2016.

[165] The Cobalt Development Institute. (2016). *Cobalt News*.

[166] Jaskula, B.W. (2016). 'Lithium'. *Miner Commod Summ US Geol Surv*. 2016;(703):100-101.

[167] Vikstrom, H., S. Davidsson, M. Hook. (2013). 'Lithium availability and future production outlooks'. *Appl Energy*. 2013;110:252-266. DOI:10.1016/j.apenergy.2013.04.005.

[168] Shedd, K.B. (2016). *USGS Minerals Commodity Summaries: Cobalt*. DOI:10.1017/CBO9781107415324.004.

[169] Umicore. (2015). 'Umicore set to invest in cobalt refining and recycling plant in Olen, Belgium'. *Umicore Press Release*.

http://www.umicore.com/en/media/press/20150930oleninve
stmentsen/. Published September 30, 2015.

[170] Lupi, C, M. Pasquali, A. Dell'era. (2005). 'Nickel and cobalt recycling from lithium-ion batteries by electrochemical processes'. *Waste Manag.* 2005;25(2):215-220.

[171] Randall, T. (2016). 'Wind and Solar Are Crushing Fossil Fuels'. *Bloomberg.* http://www.bloomberg.com/news/articles/2016-04-06/wind-and-solar-are-crushing-fossil-fuels. Published April 6, 2016.

[172] World Bank. (2009). Accelerating Innovation and Technology Diffusion. In: *World Development Report*; 2009:287-319. DOI:10.1596/978-0-8213-7987-5.

[173] IEA. (2016). *Energy Technology Perspectives 2016.*

[174] IEA. (2015). *Energy Technology Perspectives 2015.*

[175] Wattles, J. (2015). 'Bill Gates has pulled together a multinational band of investors to put billions into clean energy'. *CNN Money.* http://money.cnn.com/2015/11/29/news/economy/bill-gates-breakthrough-energy-coalition/. Published November 30, 2015.

[176] Total IEA, Energy P. (2015). Key trends in IEA public energy technology research , development and demonstration (RD & D) budgets. 2015:2-7.

[177] Commission E. (2004). 'Hydrogen economy: new EU hydrogen and fuel cell Quick Start initiative'. *Press Release Database.* http://europa.eu/rapid/press-release_IP-04-363_en.htm. Published March 18, 2004.

[178] European Commission. (2008). *HyWays The European Hydrogen Roadmap.*

[179] Global Wind Energy Council. (2012). *Irena-GWEC: 30 Years of Policies for Wind Energy.*

[180] Levring, P. (2016). 'Denmark Scares Off Investors After Energy Agenda is Jettisoned'. *Bloomberg.* http://www.bloomberg.com/news/articles/2016-06-05/denmark-scares-off-investors-after-energy-agenda-is-jettisoned. Published June 5, 2016.

[181] Enerdata. (2015). 'China revises solar feed-in tariffs for 2016'. *Newsrelease.* http://www.enerdata.net/enerdatauk/press-and-publication/energy-news-001/china-revises-solar-and-

wind-feed-tariffs-2016_35469.html. Published December 21, 2015.

[182] Zhang, S., X. Qin (2015). *Lessons Learned from China's Residential Tiered Electricity Pricing Reform*. Published May 2015.

[183] Chediak, M. (2015). 'Say Goodbye to Solar Power Susidies'. *Bloomberg BusinessWeek*. http://www.bloomberg.com/news/articles/2015-11-05/say-goodbye-to-solar-power-subsidies. Published November 5, 2015.

[184] Bloomberg New Energy Finance. (2015). Bloomberg New Ene Finance Summit 2015.

[185] IEA. (2015). 'Energy Subsidies'. *World Energy Outlook*. http://www.worldenergyoutlook.org/resources/energysubsidies/.

[186] Johnston, D. (2008). 'Changing fiscal landscape'. *J World Energy Law Bus*. 2008;1(1):31-54. DOI:10.1093/jwelb/jwn006.

[187] World Bank. (2015). Oil rents (% of GDP). *Statistics*. http://data.worldbank.org/indicator/NY.GDP.PETR.RT.ZS. Published 2015.

[188] EY. (2015). Global oil and gas tax guide 2015. 2015:(last accessed 01.08.2013). http://www.ey.com/GL/en/Industries/Oil---Gas/2013-global-oil-and-gas-tax-guide.

[189] Hirsch, J. (2015). 'Elon Musk's growing empire is fueled by $4.9 billion in government subsidies'. *Los Angeles Times*. http://www.latimes.com/business/la-fi-hy-musk-subsidies-20150531-story.html. Published May 30, 2015.

[190] Hirsch, J. (2015). 'Elon Musk: "If I cared about subsidies, I would have entered the oil and gas industry"'. *Los Angeles Times*. http://www.latimes.com/business/autos/la-fi-hy-musk-subsidies-20150602-story.html. Published June 2, 2015.

[191] Sturm, R., A.I. Njagi, L. Blyth et al. (2016). *Off-Grid Solar Market Trends Report 2016*.

[192] Aglionby, J. (2016). 'Lightbulb moment for M-Kopa'. *The Financial Times*. https://next.ft.com/content/ccfaa1ba-d0f1-11e5-831d-09f7778e7377. Published March 17, 2016.

[193] Bloomberg New Energy Finance. (2016). *BBOXX Advances Strategy with Investment from Engie*.

[194] TESLA Motors. (2010). 'Strategic Partnership: Daimler acquires stake in Tesla'. *Press Release*. April 20, 2010.

[195] Meza, E. (2016). 'Tesla partners with Hanergy on first Chinese charging stations'. *PV Magazine*. http://www.pv-magazine.com/archive/articles/beitrag/tesla-partners-with-hanergy-on-first-chinese-charging-stations_100014889/#axzz4MK67AUnX.

[196] Parrott, G. (2015). 'Tesla's First Solar-Powered Supercharger-Store-Service Center Now Open'. *Green Car Reports*. http://www.greencarreports.com/news/1096926_teslas-first-solar-powered-supercharger-store-service-center-is-almost-ready. Published March 2, 2015.

[197] Mahapatra, S. (2016). 'Lowest-Ever Solar Price Bid (2.42¢/kWh) Dropped In Abu Dhabi By JinkoSolar & Marubeni Score'. *Clean Technicachnica*. http://cleantechnica.com/2016/09/20/lowest-ever-solar-price-bid-2-42%C2%A2kwh-dropped-abu-dhabi-jinkosolar-marubeni-score/. Published September 20, 2016.

[198] Ma J., C. Trudell (2016). 'Fraud Leaves China's Electric Car Demand in Doubt'. *Bloomberg*. http://www.bloomberg.com/news/articles/2016-04-21/china-s-electric-car-subsidy-fraud-casts-doubt-on-surging-demand. Published April 21, 2016.

[199] Kariuki, J. (2016). 'M-Pesa success sets pace for mobile money uptake'. *Daily Nation*. http://www.nation.co.ke/business/M-Pesa-success-sets-pace-for-mobile-money-uptake/-/996/3257134/-/jweafr/-/index.html. Published June 20, 2016.

References Chapter 2

[1] Smil, V. (2005). *Creating the Twentieth Century: Technical Innovations of 1867-1914 and Their Lasting Impact* (1st ed.). Oxford University Press.

[2] Gordon, R.J. (2016). *The Rise and Fall of American Growth* (1st ed.). Princeton University Press.

[3] DfT. (2014). 'Transport Statistics Great Britain 2014'. *Statistics (Ber)*. 1-288. DOI:ISBN: 9780115530951. Publishes November 2014.

[4]   Townsend, S.C. (2013). 'The "Miracle" of Car Ownership in Japan's "Era of High Growth"', 1955-73. *Bus Hist.* 2013;55(3):498-523. DOI:10.1080/00076791.2013.771336.

[5]   Smil, V. (2007). 'Prime movers of globalization: The history and impact of diesel engines and gas turbines'. *J Glob Hist.* 2007;(2):373-394. DOI:10.1017/S1740022807002331.

[6]   Hellenic Shipping News. (2014). Too many ships in the world merchant fleet. http://www.hellenicshippingnews.com/984ef639-7f94-4d62-88a9-f80b3ecc6fb9/. Published April 8, 2014.

[7]   Boeing Survey. (2016). Size of aircraft fleets by region worldwide in 2014 and 2034 (in units). *Statistita.* http://www.statista.com.iclibezp1.cc.ic.ac.uk/statistics/262971/aircraft-fleets-by-region-worldwide/.

[8]   Air Transport Action Group. (2016). ATAG Facts & Figures. http://www.atag.org/facts-and-figures.html.

[9]   Sokolov, V.A. (2002). *Petroleum.* University Press of the Pacific.

[10]   Gazprom. (2014). *50th Anniversary of Underground Gas Storage in Russia.*

[11]   Agency IE. (2015). *Energy Balances of OECD Countries.* http://www.iea.org/bookshop/661-Energy_Balances_of_OECD_Countries.

[12]   Gallery, T.C. (2016). Sacrophagus. http://chernobylgallery.com/chernobyl-disaster/sarcophagus/. Accessed May 16, 2016.

[13]   Nixon, R. (1973). 128 - Special Message to the Congress on Energy Policy. The American Presidency Project. http://www.presidency.ucsb.edu/ws/?pid=3817. Published 1973. Accessed May 16, 2016.

[14]   Jamasmie, C. (2016). 'Global oil supply could face 4.5 million barrels shortfall — Wood Mackenzie'. *Mining.com.* April 27, 2016.

[15]   Smith, L. (2016). *Conventional Discoveries Outside North America Continue Their Decline ; 2015 Marked Lowest Year for Discovered Oil and Gas Volumes since 1952.*

[16]   Tang, X., M. Hook. (2015). 'Offshore oil: Investigating production parameters of fields of varying size, location and

water depth'. *Fuel*. 2015;139(January):430-440. DOI:10.1016/j.fuel.2014.09.012.

[17] Hook, M., R. Hirsch, K. Aleklett. (2009). 'Giant oil field decline rates and their influence on world oil production'. *Energy Policy*. 2009;37(6):2262-2272. DOI:10.1016/j.enpol.2009.02.020.

[18] *Coal-Cutting Machine*. (2010). (3rd ed.). The Gale Group http://encyclopedia2.thefreedictionary.com/Coal-Cutting+Machine.

[19] Jones, C.F. (2014). *Routes of Power: Energy and Modern America* (1st ed.). Harvard University Press.

[20] Broadberry, S.N. (2005). *The Productivity Race: British Manufacturing in International Perspective, 1850-1990* (1st ed.). Cambridge University Press.

[21] Ramey, V.A. (2008). *Time Spent in Home Production in the 20th Century: New Estimates from Old Data*. Vol 22. DOI:10.1017/CBO9781107415324.004.

[22] IEA. (2015). *Energy Balances of Non-OECD Countries*.

[23] IEA. (2015). *Energy Balances of OECD Countries*.

[24] Vance, B. (2013). *Coal Train Fact Check*.

[25] Freinkel, S. (2011). 'A brief history of plastic consequest of the world'. *Scientific American*. http://www.scientificamerican.com/article/a-brief-history-of-plastic-world-conquest/. Published May 29, 2011.

[26] British workers general strike to support mine workers, 1926. (2012). *Global Nonviolent Action Database*. http://nvdatabase.swarthmore.edu/content/british-workers-general-strike-support-mine-workers-1926.

[27] Mitchell, T. (2011). *Carbon Democracy: Political Power in the Age of Oil*. Verso.

[28] Campbell, C.J. (2013). *Campbell ' S Atlas of Oil and Gas Depletion* (1st ed.). Springer.

[29] Capital BMO, Energy M. (2015). *Deepwater Africa*. Published 2015.

[30] Chakhmakhchev, A., P. Rushworth. (2010). 'Global Overview of Recent Exploration Investment in Deepwater-New Discoveries, Plays and Exploration Potential'. *AAPG Search Discov Artic*. 2010;40656.

[31] Clint, O., S.E. Asia, S. Asia. (2012). *Bernstein Energy Finding Petroleum : The Significance of Deep Water to Global Supply Industry appetite for exploration continues to remain high.* Published September 2012.

[32] Daly, M. (2013). 'Future Trends in Global Oil and Gas Exploration'. *Imp Coll Centen Conf.*:1-25. http://www.bp.com/content/dam/bp/pdf/speeches/2013/Future_Trends_in_Global_Oil_and_Gas_Exploration_.pdf.

[33] Haatvedt, G.G. (2014). 'Exciting Exploration Plans NCS 2014 Statoil Leading Conventional Discovered Volume NCS Discoveries in Top League'.

[34] Janeiro, R. de. (2014). 'Oil & Gas Exploration Trends and Company Performance'. Published August, 2014.

[35] Sandrea, I., R. Sandrea. (2007). 'Global Offshore Oil: Geological Setting of Producing Provinces, E&P Trends, URR, and Medium Term Supply Outlook'. *Oil Gas J.* Published January, 2007.

[36] Editorial Staff. (2008). 'Brazil Officials Clash over Financial Crisis' Impact on E&P'. *Oil & Gas Journal.* Published November 24, 2008.

[37] Libra Oil Field, Santos Basin, Brazil. (2014). *Offshoretechnology.com.* http://www.offshore-technology.com/projects/libra-oil-field-santos-basin/.

[38] Petrobras. (2015). *Petrobras Update.*

[39] Wood Mackenzie. (2015). 'Seismic Shifts in the Oil and Gas Business'. *Wood Mackenzie.* Published October 7, 2015.

[40] Junqueira, A., J. Ozenda, A. Cardona. (2015). 'Petrobras Mapping the Production Curve'. *BTGPactual Equity Research.* Published December 17, 2015:9.

[41] United States Geological Survey. (2012). 'An Estimate of Undiscovered Conventional Oil and Gas Resources'. *USGS Fact Sheet.* 2012;2012–3042(March):6. http://pubs.usgs.gov/fs/2012/3042/.

[42] Perlin, J. (2013). *Let It Shine: The 6000 Year Story of Solar Energy.* New World Library.

[43] Jones, G., L. Bouamane. (2012). 'Power from Sunshine: A Business History of Solar Energy'. *Harvard Bus Sch.*:86.

[44] Carter, J. (1979). 'Solar Energy Message to the Congress'. *The American Presidency project*. http://www.presidency.ucsb. edu/ws/?pid=32503. Published June 20, 1979.

[45] Kimura, O. (2006). '30 Years of Solar Energy Development in Japan: Co-Evolution Process of Technology, Policies, and the Market'. *Power*. 2006;(November):2-11.

[46] REN21. (2008). *Renewables 2007 Global Status Report*.:1-54. http://www.ren21.net/pdf/RE2007_Global_Status_Report.pdf.

[47] The Editors. (2013). 'Obama Administration Becomes the Third to Install Solar Panels on White House Grounds'. *ThinkProgress*. https://thinkprogress.org/obama-adminis-tration-becomes-the-third-to-install-solar-panels-on-white-house-grounds-8afd1867cd0f#.18myvbebe. Published August 15, 2013.

[48] The Editors. (2016). 'Residential Solar PV Price Index'. *Solar Choice*. http://www.solarchoice.net.au/blog/news/residen-tial-solar-pv-prices-august-2016-080816. Published August 8, 2016.

[49] Finke, S. (2016). 'Photovoltaik-Kosten'. *PVS SolarStrom*. http://www.photovoltaiksolarstrom.de/photovoltaik-kosten. Published August, 2016.

[50] Wirth, H. (2016). *Aktuelle Fakten Zur Photovoltaik in Deutschland*. https://www.ise.fraunhofer.de/de/veroef-fentlichungen/veroeffentlichungen-pdf-dateien/studien-und-konzeptpapiere/aktuelle-fakten-zur-photovoltaik-in-deutschland.pdf.

[51] Kimura, K., R. Zissler. (2016). 'Comparing Prices and Costs of Solar PV in Japan and Germany: The Reasons Why Solar PV is more Expensive in Japan'. *Renew Energy Inst*. 2016;(March):25.

[52] Australian Government Department of Industry Innovation and Science. (2016). *Energy in Australia 2015*. DOI:10.1049/ep.1987.0167.

[53] BSW-Solar. (2016). 'Solarspeicher-Preise um ein Drittel Gefallen'. *BSW-Solar*. https://www.solarwirtschaft.de/presse/pressemeldungen/pressemeldungen-im-detail/news/solars-peicher-preise-um-ein-drittel-gefallen.html. Published June 21, 2016.

[54] Michel, J. (2016). 'Germany Sets a New Solar Storage Record'. *Energy Post*. http://www.theenergycollective.com/energy-post/2383073/germany-sets-a-new-solar-storage-record. Published July 18, 2016.

[55] Blackman, J. (2016). 'German Storage Market to Hit USD 1 Billion by 2021 on Falling Costs and Subsidies'. *PVTECH*. http://www.pv-tech.org/news/german-storage-market-to-hit-1-billion-by-2021. Published July 28, 2016.

[56] Morris, C. (2015). 'Grid Defection and Why We Don't Want It'. *Energy Transit Ger Energiewende*. Published July, 2015. http://energytransition.de/2015/06/grid-defection/.

[57] Sturm, R., A.I. Njagi, L. Blyth, et. al. (2016). *Off-Grid Solar Market Trends Report 2016*.

[58] Kariuki, J. (2016). 'M-Pesa Success Sets Pace for Mobile Money Uptake'. *Daily Nation*. http://www.nation.co.ke/business/M-Pesa-success-sets-pace-for-mobile-money-uptake/-/996/3257134/-/jweafr/-/index.html. Published June 20, 2016.

[59] Aglionby, J. (2016). 'Lightbulb Moment for M-Kopa'. *The Financial Times*. https://next.ft.com/content/ccfaa1ba-d0f1-11e5-831d-09f7778e7377. Published March 17, 2016.

[60] Sharma, A. (2014). 'Modi Government Working to Illuminate 25000 Villages Through Micro-Grids'. *One India*. http://www.oneindia.com/india/modi-govt-working-illuminate-25000-villages-through-micro-grids-1534736.html. Published October 4, 2014.

[61] Mukherjee, B. (2015). 'Boond Makes Every Drop Count When it Comes to Lighting Rural India'. *Your Story*. Published July, 2015. http://social.yourstory.com/2015/07/boond/.

[62] Wood, E. (2016). 'Social Enterprise Co. Plans to Install 4,000 Solar Microgrids in India'. *MicroGrid Knowledge*. https://microgridknowledge.com/solar-microgrids-in-india/.

[63] Hansen, M. (2015). 'Wind Costs More Than You Think Due to Massive Federal Subsidies'. *The Conversation*. http://theconversation.com/wind-costs-more-than-you-think-due-to-massive-federal-subsidies-38804.

[64] Stop These Things. (2014). 'Professor Ross McKitrick: Wind Turbines Don't Run on Wind, They Run on Subsidies'. *Stop*

These Things. https://stopthesethings.com/2014/08/31/
professor-ross-mckitrick-wind-turbines-dont-run-on-wind-
they-run-on-subsidies/.

[65] Kelly-Gagnon. (2015). 'Quebec Should Ditch Wind Power
Subsidies and Go for Oil'. *Huffington Post Blog.* http://www.
huffingtonpost.ca/michel-kellygagnon/energy-choices-
quebec_b_7156344.html. Published April 28, 2015.

[66] Shahan, Z. (2014). 'History of Wind Turbines'. *Renewable
Energy World.* http://www.renewableenergyworld.com/
ugc/blogs/2014/11/history-of-wind-turbines.html. Published
November 21, 2014.

[67] Neij, L., P.D. Andersen, M. Durstewitz, P. Helby, M. Hoppe-
Kilper, P-E. Morthorst. (2003). *Experience Curves: A Tool for
Energy Policy Assessment.* DOI:10.1016/S0960-0779(02)00434-
4.

[68] Vestas Wind Systems. (2006). Life Cycle Assessment of
Electricity Produced from Onshore Sited Wind Power Plants
Based on Vestas V82-1.65 MW Turbines'.:1-77. https://www.
vestas.com/~/media/vestas/about/sustainability/pdfs/lca
v82165 mw onshore2007.ashx.

[69] D'Souza, N., E. Gbegbaje-Das, P. Shonfield. (2011). 'Life
Cycle Assessment of Electricity Production from a Vestas
V112 Turbine Wind Plant'. *Production.* DOI:10.1016/S0140-
6701(01)80378-7.

[70] GAMESA. (2013). *Environmental Product Declaration European G114-2.0MW On-shore Wind Farm.*:38.

[71] Eisen, J.B. (2010). 'China's Renewable Energy Law: A Platform
for Green Leadership?'. *William Mary Environ Law Policy Rev.*
2010;1(1):0-52.

[72] Liu, Z., W. Zhang, C. Zhao, J. Yuan. (2015). 'The Economics
of Wind Power in China and Policy Implications'. *Energies.*
2015;8(2):1529-1546. DOI:10.3390/en8021529.

[73] Jianxiang, Y. (2016). 'China Reduces FITs Over Two-Year Period'. *Wind Power Mon.* Published January, 2016. http://www.
windpowermonthly.com/article/1378719/china-reduces-fits-
two-year-period.

[74]   Energy. GOV. (2016). *Renewable Electricity Production Tax Credit*. http://energy.gov/savings/renewable-electricity-production-tax-credit-ptc.

[75]   Oxera. (2014). *Almost a Reform: The New German Support Scheme for Renewable Electricity*. http://www.oxera.com/Latest-Thinking/Agenda/2014/Almost-a-reform-the-new-German-support-scheme-for.aspx.

[76]   Appunn, K. (2016). 'EEG Reform 2016: Switching to Auctions for Renewables'. *Clean Energy Wire*. July 2016:3. https://www.cleanenergywire.org/factsheets/eeg-reform-2016-switching-auctions-renewables.

[77]   IRENA. (2016). *Wind Power: Technology Brief*. DOI:10.1049/ep.1976.0231.

[78]   Bolinger, M., R. Wiser. (2011). 'Understanding Trends in Wind Turbine Prices Over the Past Decade'. *Energy Policy*. 2011;42(October):628-641. http://linkinghub.elsevier.com/retrieve/pii/S0301421511010421.

[79]   Parkinson, G. (2016). 'New Low for Wind Energy Costs: Morocco Tender Averages USD30/MWh'. *REneweconomy*. http://reneweconomy.com.au/2016/new-low-for-wind-energy-costs-morocco-tender-averages-us30mwh-81108.

[80]   ReNews. (2014). 'Quebec 450 MW RFP Winners'. *reNews*. http://renews.biz/81179/quebec-names-450mw-rfp-winners/.

[81]   Sterling, T. (2016). 'Denmark's Dong Energy Wins Dutch Offshore Wind Tender'. *Mail Online*. http://www.dailymail.co.uk/wires/reuters/article-3675746/Denmarks-DONG-Energy-wins-Dutch-offshore-wind-tender.html. Published July 5, 2016.

[82]   ReNews. *Vattenfall Wins Horns Rev 3 Tender*. http://renews.biz/84884/vattenfall-wins-horns-rev-3-tender/.

[83]   Vattenfall. (2016). 'Vattenfall Wins Danish Near Shore Wind Tender'. *Press Release*. https://corporate.vattenfall.com/press-and-media/press-releases/2016/vattenfall-wins-danish-near-shore-wind-tender/. Published December 9, 2016.

[84]   HMSQ, E. (2000). *Innovation From Coal to Oil*.

[85]   John, M. (2008). 'Extraordinary Foresight Made Winston Churchill Great'. *The Telegraph*. http://www.telegraph.co.uk/culture/books/3671962/John-McCain-Extraordinary-fore-

sight-made-Winston-Churchill-great.html. Published March 20, 2008.

[86] Leloux, J., L.N. Fernandez, D. Trebosc. (2011). *Performance Analysis of 10,000 Residential PV Systems in France and Belgium.* http://oa.upm.es/12617/. Accessed December 27, 2014.

[87] Jordan, D., S. Kurtz. (2013). 'Photovoltaic Degradation Rates: An Analytical Review'. *Prog Photovoltaics Res Appl.* 2013;21:12-29. DOI:10.1002/pip.

[88] Colthorpe, A. (2015). 'Sonnenbatterie Adds "Very Long Lifetime" to Residential Battery Systems'. *Energy Storage.* http://www.energy-storage.news/news/sonnenbatterie-adds-very-long-lifetime-to-residential-battery-systems. Published April 10, 2015.

[89] Bradley, R. (2016). 'From Zond to Enron Wind to GE Wind: Founder Interview'. *MasterResource.* https://www.master-resource.org/windpower/history-us-wind-industry-zond-enron-ge/. Published May 4, 2016.

[90] Clark, P. (2016). 'Dong Energy in Offshore Wind Cost Breakthrough'. *The Financial Times.* https://next.ft.com/content/18b0f2b6-42db-11e6-b22f-79eb4891c97d. Published July 5, 2016.

References Chapter 3

[1]   The Economist Staff. (1999). 'Standard ogre'. *The Economist.* Published December 23, 1999.

[2]   Engdahl, W.F. (2004). *A Century of War: Anglo-American Oil Politics and the New World Order* (2nd ed.). Pluto Press.

[3]   Morton, M.Q. (2013). *Buraimi: The Struggle for Power, Inluence and Oil in Arabia* (1st ed.). IB Tauris & Co.

[4]   Kinzer S. (2007). *Overthrow: America's Century of Regime Change from Hawaii to Iraq* (2nd ed.). Times Books.

[5]   Kinzer, S. (2014). *The Brothers: John Foster Dulles, Allen Dulles, and Their Secret World War* (2nd ed.). St. Martin's Griffin.

[6]   Morris, R. (2003). 'A tyrant 40 years in the making'. *The New York Times.* http://www.nytimes.com/2003/03/14/opinion/a-

tyrant-40-years-in-the-making.html. Published March 14, 2003.

[7]   Mark CR. (2005). *CRS Issue Brief for Congress Israel: U.S. Foreign Assistance.*

[8]   BP. (2016). BP Statistical Review of World Energy June 2016. 2016. http://www.bp.com/statisticalreview.

[9]   Frankel G. (2004). U.S. Mulled Seizing Oil Fields in '73. *The Washington Post.* https://www.washingtonpost.com/archive/politics/2004/01/01/us-mulled-seizing-oil-fields-in-73/0661ef3e-027e-4758-9c41-90a40bbcfc4d/. Published January 1, 2004.

[10] Dreyfuss, R. (2005). *Devil's Game: How the United States Helped Unleash Fundamentalist Islam.* Metropolitan Press

[11] Sharp, J.M. (2016). *Congressional Research Service Egypt: Background and US Relations.*

[12] Mitchell, T. (2011). *Carbon Democracy: Political Power in the Age of Oil.* Verso

[13] Simons G. (1998). *Saudi Arabia: The Shape of a Client Fuedalism.* Palgrave Macmillan

[14] BBC News. (1999). Arms sales fuel BAe's profits. *BBC News.* http://news.bbc.co.uk/2/hi/business/285963.stm. Published February 25, 1999.

[15] Farmer B. (2016). Margaret Thatcher stopped Al-Yamamah arms deal going to French with secret personal "tactical arguments" meeting with Saudis. *The Daily Telegraph.* http://www.telegraph.co.uk/news/2016/08/23/margaret-thatcher-stopped-al-yamamah-arms-deal-going-to-french-w/. Published August 24, 2016.

[16] Mozgovaya N. (2010). U.S. announces 60 billion dollar Arms Sale to Saudi Arabia, Says Israel Doesn't Object. *Israel News.* http://www.haaretz.com/israel-news/u-s-announces-60b-arms-sale-to-saudi-arabia-says-israel-doesn-t-object-1.320307. Published October 20, 2010.

[17] Black I, Tisdall S. (2010). Saudi Arabia urges US attacks on Iran to stop nuclear programme. *The Guardian.* https://www.theguardian.com/world/2010/nov/28/us-embassy-cables-saudis-iran. Published November 28, 2010.

[18]   London Institute of Petroleum. (1999). Full text of Dick Cheney's speech at the IP Autumn lunch.

[19]   Sciolino E, Tyler PE. (2001). Some Pentagon Officials and Advisers Seek to Oust Iraq's Leader in War's next Phase. *The New York Times*. http://www.nytimes.com/2001/10/12/international/middleeast/12EXPA.html. Published October 12, 2001.

[20]   GPO. (2001). Defense Policy Board Advisory Group; Meeting. *Fed Regist*. 2001;66(175):47021.

[21]   Democracy Now. (2007). *General Wesley Clark Weighs Presidential Bid: "I Think About It Every Day."* Democracy Now http://www.democracynow.org/shows/2007/3/2?autostart=true.

[22]   Cutler RJ. (2013). *The World According to Dick Cheney*. USA: Showtime Networks

[23]   Al Jazeera Staff. (2011). Colin Powell regrets Iraq War Intelligence. *Al Jazeera*. http://www.aljazeera.com/news/americas/2011/09/20119116916873488.html. Published September 11, 2011.

[24]   Clarke WR. (2005). *Petrodollar Warfare: Oil, Iraq and the Future of the Dollar*. 1st ed. New Society Publishers

[25]   Unger C. (2004). *House of Bush, House of Saud: The Hidden Relationship between the World's Two Most Powerful Dynasties*. Gibson Square

[26]   Kagan D, Schmitt G, Donnely T. (2000). *Rebuilding America's Defenses: Strategy, Forces and Resources For a New Century*. http://www.informationclearinghouse.info/pdf/RebuildingAmericasDefenses.pdf.

[27]   Bookman J. (2002). The Presiden'ts Real Goal in Iraq. *Atlanta Journal Constitution*. http://www.informationclearinghouse.info/article2319.htm. Published September 29, 2002.

[28]   Hayes TC. (1990). CONFRONTATION IN THE GULF; The Oilfield Lying Below the Iraq-Kuwait Dispute. *International New York Times*. http://www.nytimes.com/1990/09/03/world/confrontation-in-the-gulf-the-oilfield-lying-below-the-iraq-kuwait-dispute.html?pagewanted=all&src=pm. Published September 3, 1990.

[29] Editorial Staff. (1990). Excerpts from Iraqi Document on Meeting with U.S. Envoy. *International New York Times*. Excerpts From Iraqi Document on Meeting with U.S. Envoy. Published September 23, 1990.

[30] Gordon MR. (1990). Bush sends U.S. Force to Saudi Arabia as Kingdom Agrees to Confront Iraq; Bush's Aims: Deter Attack, Send a Signal. *International New York Times*. http://www.nytimes.com/1990/08/08/world/bush-sends-us-force-saudi-arabia-kingdom-agrees-confront-iraq-bush-s-aim-s-deter.html?pagewanted=all. Published August 8, 1990.

[31] United Nations Security Council. (1990). Resolution 678. 1990.

[32] Carlisle PR, Bowman JS. (2010). *Persian Gul War*. 2nd ed. Chelsea House Publishers

[33] Pope C. (2004). Cheney Changed his view on Iraq: He said in '92 Saddam not worth U.S. casualties. *Seattle Post*. http://www.seattlepi.com/national/article/Cheney-changed-his-view-on-Iraq-1155325.php. Published September 28, 2004.

[34] Wong A. (2016). The untold story behind Saudi Arabias 41 year U.S. Debt Secret. *Bloomberg*. http://www.bloomberg.com/news/features/2016-05-30/the-untold-story-behind-saudi-arabia-s-41-year-u-s-debt-secret. Published May 31, 2016.

[35] Islam F. (2003). Iraq nets handsome profit by dumping dollar for euro. *The Guardian*. https://www.theguardian.com/business/2003/feb/16/iraq.theeuro. Published February 16, 2003.

[36] White G. (2012). Iran presses ahead with dollar attack. *The Daily Telegraph*. http://www.telegraph.co.uk/finance/commodities/9077600/Iran-presses-ahead-with-dollar-attack.html. Published February 12, 2012.

[37] Hoff B. (2016). Hillary Emails reveal true motive for Libya Intervention. *Foreign Policy Journal*. http://www.foreignpolicyjournal.com/2016/01/06/new-hillary-emails-reveal-true-motive-for-libya-intervention/. Published January 6, 2016.

[38] Nahmias R. (2006). Syria switches to the Euro due to US sanctions. *Ynet.news*. http://www.ynetnews.com/articles/0,7340,L-3322316,00.html. Published November 1, 2006.

[39] Sputnik News. (2015). Russia reduces US Treasury Holdings by 40 percent. *Website*. https://sputniknews.com/business/20150616/1023419663.html. Published June 16, 2015.

[40] International Reserves of the Russian Federation. (2016). *Bank of Russia*. https://www.cbr.ru/eng/hd_base/Default.aspx?Prtid=mrrf_m. Published 2016.

[41] Engdahl WF. (2016). Why are Russia and China Buying Gold, Tons of it? *GlobalResearch*. http://www.globalresearch.ca/why-are-russia-and-china-buying-gold-tons-of-it/5518896. Published March 30, 2016.

[42] Wong A. (2016). As China Dumps Treasures, World Sees No Better Place for Refuge. *Bloomberg*. http://www.bloomberg.com/news/articles/2016-01-10/china-retreat-from-u-s-bonds-prompts-shrugs-where-fear-reigned. Published January 10, 2016.

[43] Reagan M. (2014). Putin will do what he wants in ukraine. *NewsMax*. http://www.newsmax.com/MichaelReagan/Putin-Ukraine-oil-Gorbachev/2014/03/07/id/556690/. Published March 7, 2014.

[44] Woodward B, Ottaway DB. (1987). President, Saudis Met Twice; Funds Flowed to Contras After Talks. *The Washington Post*. http://www-personal.umich.edu/~jrcole/qaeda/fahdreagan.htm. Published 1987.

[45] Byron C. (1981). Problems for Oil Producers. *TIME Magazine*. June 22, 1981.

[46] Reynolds DB, Kolodziej M. (2008). Former Soviet Union oil production and GDP decline: Granger causality and the multi-cycle Hubbert curve. *Energy Econ*. 2008;30(2):271-289. DOI:10.1016/j.eneco.2006.05.021.

[47] Buckley N, Arnold M. (2016). Herman Gref, Sberbank's modernising sanctions sruvivor. *The Financial Times*. https://www.ft.com/content/4abbcba6-c413-11e5-808f-8231cd71622e. Published January 31, 2016.

[48] Editorial Staff. (2015). FIFA crisis: US charges 16 more officials after earlier Zurich arrests. *BBC Sport*. http://www.bbc.com/sport/football/34991874. Published December 4, 2015.

[49] Dogan YP. (2015). Expert: Oil Price Wars Fatally Wounded the Petrodollar. *Russia Insider*. http://russia-insider.com/ru/2015/02/04/3126. Published February 15, 2015.

[50] Zengerle P, Baum B. (2015). U.S. Senators urge FIFA not to let Russia host World Cup 2018. *Reuters*. http://www.reuters.com/article/us-soccer-fifa-congress-idUSK-BN0MS52G20150401. Published April 1, 2015.

[51] Zarate, J.C. (2013). 'Treasury's War: The Unleashing of a New Era of Financial Warfare'. *Public Affairs.*

[52] Stern, J. (2006). 'The Russian-Ukrainian Gas Crisis of January 2006'. *Oxford Inst Energy Stud.* 2006;(October 2005):17. http://www.avim.org.tr/icerik/energy-gas.pdf.

[53] Pirani, S., J. Stern, K. Yafimava. (2009). *The Russo-Ukrainian Gas Dispute of January 2009: A Comprehensive Assessment.* Vol NG 27. http://www.oxfordenergy.org/wpcms/wp-content/uploads/2010/11/NG27-TheRussoUkrainianGasDispute-ofJanuary2009AComprehensiveAssessment-JonathanStern-SimonPiraniKatjaYafimava-2009.pdf.

[54] Pirani, S., K. Yafimava. (2016). *Russian Gas Transit Across Ukraine Post-2019: Pipeline Scenarios, Gas Flow Consequences, and Regulatory Constraints.* https://www.oxfordenergy.org/wpcms/wp-content/uploads/2016/02/Russian-Gas-Transit-Across-Ukraine-Post-2019-NG-105.pdf.

[55] Union, E. (2013). 'Russian Gas Imports To Europe and Security of Supply'. *Factsheet.*:5-7.

[56] Zhdannikov, D., D. Pinchuk. (2016). 'Gazprom Warns of Steep Gas Transit Cuts Via Ukraine After 2020'. *Reuters.* http://www.reuters.com/article/us-gazprom-exports-ukraine-idUSKCN0Z20YR. Published June 16, 2016.

[57] Gazprom, BASF, E.ON, Engie, OMV and Shell Sign Shareholders Agreement on the Nord Stream 2 Project. (2015). *Nord Stream 2.* https://www.nord-stream2.com/media-info/news/gazprom-basf-e-on-engie-omv-and-shell-sign-shareholders-agreement-on-the-nord-stream-2-project-2/. Published September 4, 2015.

[58] Slowikowski, M. (2016). 'Is Nord Stream 2 Still a Good Deal for Europe?'. *Oilprice.com.* http://oilprice.com/Energy/Nat-

ural-Gas/Is-Nord-Stream-2-Still-A-Good-Deal-For-Europe. html. Published August 30, 2016.

[59] Foy, H., J. Farchy. (2016). 'Nord Stream 2 Pipeline Risks Delays Due to Polish Hurdle'. *The Financial Times*. https://www. ft.com/content/e2cf7602-5411-11e6-befd-2fc0c26b3c60. Published July 28, 2016.

[60] Stern, J., S. Pirani, K. Yafimava. (2015). 'Does the Cancellation of South Stream Signal a Fundamental Reorientation of Russian Gas Export Policy?'. *Oxford Energy Comment*. 2015;(January):15.

[61] The Jamestown Foundation. (2016). 'Post Coup: Gazprom Still Eager to Complete Turkish Stream'. *Oilprice.com*. http://oilprice.com/Energy/Energy-General/Post-Coup-Gazprom-Still-Eager-To-Complete-Turkish-Stream.html. Published August 5, 2016.

[62] RT. (2016). 'Ankara to Share Cost of Turkish Stream Pipeline, Says Erdogan'. *RT*. https://www.rt.com/business/355549-turkish-stream-cost-erdogan/. Published August 11, 2016.

[63] Geropoulos, K. (2016). 'Russia Pushes Tesla Pipeline Through Balkans'. *New Europe*. https://www.neweurope.eu/article/russia-pushes-tesla-pipeline-through-balkans/. Published August 20, 2016.

[64] Cedigaz. (2015). 'Global Growth Rates: Macroeconomic Indicators'. *Cedigaz Reference Scenario*. Published February, 2015. http://www.cedigaz.org/documents/2015/CEDI-GAZProspects2015.pdf.

[65] Matthes, O. (2014). 'Putin's "Last and Best Weapon" Against Europe: Gas'. *Newsweek*. Published September 24, 2014.

[66] NAM. (2016). *Productiecijfers NAM 2015*. *Webpage*. http://www.nam.nl/nl/news/news-archive-2016/production-figures-nam-2015.html. Published January 7, 2016.

References Chapter 4

[1] Hubbert, M.K. (1956). 'Nuclear Energy and the Fossil Fuels'. *Drill Prod Pract*. 1956;(95):57.

[2]   Inman, M. (2016). *The Oracle of Oil: A Maverick Geologist's Quest for a Sustainable Future*. W.W. Norton & Company.

[3]   U.S. Energy Information Administration. (2016). US *Crude Oil Production*. https://www.eia.gov/dnav/pet/pet_crd_cr-pdn_adc_mbblpd_m.htm.

[4]   Hirsch, R.L., R. Bezdek, R. Wendling. (2005). *Peaking of World Oil Production, Impacts, Mitigation & Risk Management*.

[5]   U.S Energy Information Administration. (2016). *International Energy Outlook 2016*. http://www.eia.gov/forecasts/ieo/world.cfm.

[6]   IEA. *Oil Market Report*.

[7]   Robertson, S., R. Westwood. (2013). *Global Offshore Prospects*. 2013;(26 September).

[8]   Rystad Energy. (2016). 'Rustad Energy UCube'. *Database*. http://www.rystadenergy.com/Products/EnP-Solutions/UCube.

[9]   Statistics Canada. (2016). 'Historical Supply and Disposition of Crude Oil and Equivalent, Monthly'. *Database*. http://www5.statcan.gc.ca/cansim/a34?lang=eng&mode=tableSummary&id=1260001&p2=9.

[10] Teslik, L.H. (2008). 'Royal Dutch Shell CEO on the End of "Easy Oil"'. *Council on Foreign Relations*. http://www.cfr.org/oil/royal-dutch-shell-ceo-end-easy-oil/p15923. Published April 7, 2008.

[11] IEA. (2006). *World Energy Outlook 2006*.

[12] IEA. (2015). *World Energy Outlook 2015*.

[13] IEA. (2015). *Energy Balances of Non-OECD Countries*.

[14] IEA. (2015). *Energy Balances of OECD Countries*.

[15] Höök, M., S. Davidsson, S. Johansson, X. Tang. (2014). 'Decline and Depletion Rates of Oil Production: A Comprehensive Investigation'. *Philos Trans A Math Phys Eng Sci*. 2014;372(2006):20120448. DOI:10.1098/rsta.2012.0448.

[16] IEA. (2008). *World Energy Outlook 2008*.

[17] Production GO. Finding the Critical Numbers What Are the Real Decline Rates for Global Oil Production ?

[18] Wood Mackenzie. (2015). 'Seismic Shifts in the Oil and Gas Business'. *Wood Mackenzie*. Published October 7, 2015.

[19] Bloomberg Energy Intelligence. (2016). 'Oil & Gas Industry Organic E&P'. *Expl. & DEV.*

[20] Sandrea, I.R. (2006). Global E&P Capex and Liquid Capacity Trends, and Medium Term Outlook.

[21] Chen, S., J. Yang, R. Al-Rikabi. (2016). 'China Helps Balance Oil as Aging Fields Lag Rising Refiners'. *Bloomberg News*. http://www.bloomberg.com/news/articles/2016-05-16/china-helping-balance-oil-as-thirsty-refiners-rely-on-old-fields. Published May 16, 2016.

[22] BP. (2016). *BP Statistical Review of World Energy June 2016.* http://www.bp.com/statisticalreview.

[23] Smith, L. (2016). *Conventional Discoveries Outside North America Continue Their Decline: 2015 Marked Lowest Year for Discovered Oil and Gas Volumes since 1952.*

[24] Campbell, C.J. (2013). *Campbell ' S Atlas of Oil and Gas Depletion.* (1st ed.) Springer.

[25] Waldie, J. (2015). 'Offshore Support Vessels and Macro: Economic Overview of the offshore Industry'. *Wood Mackenzie.* 2015:50.

[26] PdVSA. (2015). *Desarrollo de la Faja Petrolífera del Orinoco.* 2015:100.

[27] Katakey, R. (2016). 'Drillers Can't Replace Lost Output as $100 Oil Inheritance Spent'. *Bloomberg Business.* March 23, 2016:1-4.

[28] The Editors. (2016). 'WoodMac: Growing List of Deferred Upstream Projects Reaches'. *Oil & Gas Journal.* Published January 25, 2016.

[29] Nicolaisen LE. (2015). *Oil Market -an Update from Rystad Energy.*

[30] Piotrowski, M. (2016). 'Rystad: 70 USD Oil Needed to Stimulate Enough Supply and Meet Long-Term Demand'. *The Fuse.* http://energyfuse.org/rystad-70-oil-needed-stimulate-enough-supply-meet-long-term-demand/. Published July 12, 2016.

[31] Oil & Gas Journal. (2016). 'CAPP : Capital Spending in Canadian Upstream to Decline in 2016'. *Oil & Gas Journal.* Published April 18, 2016.

[32] Pettigrew, G. (2015). 'Major Projects and Operational Trends'. *Marine Technical Society, Houston Section*. May 28, 2015:28.

[33] BGR. (2014). Energy Study 2014: *Reserv Resour Availab Energy Resour*. 18-131.

[34] WEC. (2013). 'World Energy Resources'. *World Energy Counc Rep*. 468. DOI:http://www.worldenergy.org/wp-content/uploads/2013/09/Complete_WER_2013_Survey.pdf.

[35] Report S.I. (2015). *U.S. Geological Survey Assessment of Reserve Growth Outside of the United States*.

[36] Christopher, J., Schenk, A. Troy, R.R. Cook, Charpentier, R.M. Pollastro, R.T. Klett, M.E. Tennyson, M.A. Kirschbaum, M.E. Brownfield, J.K.P. (2009). 'An Estimate of Recoverable Heavy Oil Resources of the Orinoco Oil Belt, Venezuela'. *US Geol Surv*.

[37] Attanasi, E.D., R.F. Meyer. (2007). 'Natural Bitumen and Extra-Heavy Oil'. *Surv Energy Resour*. World Energy:119-143.

[38] U.S. Energy Information Administration. (2015). *U.S. Crude Oil and Natural Gas Proved Reserves*. https://www.eia.gov/naturalgas/crudeoilreserves/.

[39] United States Geological Survey. (2012). 'An Estimate of Undiscovered Conventional Oil and Gas Resources'. *USGS Fact Sheet*. 2012;2012–3042(March):6. http://pubs.usgs.gov/fs/2012/3042/.

[40] Schwartz, N.D. (2005). 'Chevron's Dave O'Reilly: Pumped Up'. *Fortune Magazine*. http://archive.fortune.com/magazines/fortune/fortune_archive/2005/09/05/8271386/index.htm. Published September 5, 2005.

[41] Necessarily, N. (2015). 'Oil & Gas Deepwater : In Deep Trouble? Not Necessarily'.

[42] Muggeridge, A., A. Cockin, K. Webb, et. al. (2014). 'Recovery Rates, Enhanced Oil Recovery and Technological Limits'. *Philos Trans A Math Phys Eng Sci*. 2014;372(2006):20120320. DOI:10.1098/rsta.2012.0320.

[43] IEA. (2013). *Resources to Reserves 2013 - Oil, Gas and Coal Technologies for the Energy Markets of the Future*. DOI:10.1787/9789264090705-en.

[44] Birn, K., J. Meyer. (2015). 'Oil Sands Cost and Competitiveness'.

[45] Little, A.D. (2015). 'Trends and Challenges in the Heavy Crude Oil Market'. *Heavy Oil Working Group*. Bogota. 23.

[46] Lower 48 Oil&Gas: Breakeven Analysis and Company Benchmarking Update. (2016).

[47] Capital BMO, Energy M. (2015). 'Deepwater Africa'.

[48] Stevens, P. (2010). 'The "Shale Gas Revolution": Hype and Reality'. *Chatham House Rep.* 4-20. http://www.chathamhouse.org.uk/files/17317_r_0910stevens.pdf.

[49] Nixon, R. (1973). *Special Message to the Congress on Energy Policy*. http://www.presidency.ucsb.edu/ws/?pid=3817.

[50] Shackouls, B., L.R. Raymond, M. Nichols, S, Abrahan. (2003). *Balacing Natural Gas Policy: Fueling the Demands of a Growing Economy*. DOI:10.1017/CBO9781107415324.004.

[51] Administration USEI. (2015). *Assumptions to the Annual Energy Outlook 2015 : Oil and Gas Supply Module*.

[52] Rystad Energy. (2015). 'The Oil Price is Falling But So is the Breakeven Price for Shale'. *News release*. http://www.rystad-energy.com/NewsEvents/Newsletters/UsArchive/us-q1-2015.

[53] Crooks, E. (2016). 'Cost Reductions Help U.S. Shale Oil Industry Pass First Real Test'. *Financial Times*. https://www.ft.com/content/65ebdd54-6b79-11e6-ae5b-a7cc5dd5a28c. Published August 28, 2016.

[54] The next shock? (1999). *Economist*. http://www.economist.com/node/188181.

[55] Scott, C. (2008). 'T. Boone Pickens Predicts 200 A Barrel Oil'. *CBS News*. http://www.cbsnews.com/news/t-boone-pickens-predicts-200-a-barrel-oil/. Published August 27, 2008.

[56] Maxwell, C. (2008). 'What 300-a-Barrel Oil Will Mean for You'. *Barron's*. http://www.barrons.com/articles/SB122065354946305325.

[57] Hamilton, J.D. (2011). 'Historical Oil Shocks'. *Natl Bur Econ Res*. 2011;53(9):1689-1699. DOI:10.1017/CBO9781107415324.004.

[58] Agency, I.E. (2015). *Energy Balances of OECD Countries*. http://www.iea.org/bookshop/661-Energy_Balances_of_OECD_Countries.

[59] Eurostat. (2016). 'Consumption of Energy'. *Statistics Explained*. http://ec.europa.eu/eurostat/statistics-explained/index.php/Consumption_of_energy. Published July 2016.

[60] Hurd, D., S. Park, J. Kan. (2014). *China's Coal to Olefins Industry*.

[61] Abdul-hamid, O.S., A. Odulaja, H. Hassani, H. Hafidh, A. Antini, C. Bayer. (2016). *OPEC Annual Statistical Bulletin 2016*.

[62] The European Commission Directorate General For Energy. (2016). 'EU Crude Oil Imports and Supply Cost'. *Webpage*. https://ec.europa.eu/energy/en/data-analysis/eu-crude-oil-imports.

[63] Rystad Energy. (2016). 'Global Liquids Cost Curve: An Update'. *News release*. http://www.rystadenergy.com/News-Events/PressReleases/global-liquids-cost-curve-an-update. Published April 14, 2016.

[64] *Argentina's Shale Oil and Gas: Challenges and Opportunities*.

[65] Gonzales, P. (2016) 'Oil at $77? Argentina Marches to a Different Drummer'. *Bloomberg Business*. http://www.bloomberg.com/news/articles/2015-08-25/oil-at-77-argentina-marches-to-a-different-drummer. Accessed May 14, 2016.

[66] Sputniknews. (2016). 'Russia Joins the Shale Game After Building Own Technology Thanks to Sanctions'. *Webpage*. https://sputniknews.com/russia/20160831/1044818976/russia-shale-oil-technology.html. Published August 31, 2016.

[67] U.S. Energy Information Administration. (2015). *Technically Recoverable Shale Oil and Shale Gas Resources: Russia*.

[68] Gazprom Neft. (2016). 'Hydraulic Fracturing Operations at Fields with Unconventional Hydrocarbon Reserves'. *Webpage*. http://www.gazprom-neft.com/company/business/exploration-and-production/technology/.

[69] Hunn, D. (2016). 'Mexico Could Open Shale Fields to U.S. Drillers Next Year'. *Fuelfix*. http://fuelfix.com/blog/2016/09/23/mexico-could-open-shale-fields-to-u-s-drillers-next-year/. Published September 23, 2016.

[70] Stevens, S., K. Moodhe. (2016). 'New Bid Round Accelerates Mexico's Shale Potential'. *Oil & Gas Journal*. http://www.ogj.com/articles/print/volume-114/issue-6/exploration-and-development/new-bid-round-accelerates-mexico-s-shale-potential.html. Published June 6, 2016.

[71] Campbell, C.J. (2005). *Oil Crisis*. Multi-Science Publishing.

[72] Al-Husseini, S.I. (2007). 'Long-Term Oil Supply Outlook: Constraints On Increasing Production Capacity'. *Oil & Money*: 21.

[73] Hess. (2015). 'Hess Corporation'. *UBS Global Oil and Gas Conference*: 28.

[74] Pakparvar, M. (2016). 'Energy Outlook and Opportunities in Iran'. *Institute for Political & International Studies*: 23.

[75] Blas, J. (2015). 'Libya Oilfield Write-Of by Total is Ominous Sign to other Firms'. *Bloomberg*. http://www.bloomberg.com/news/articles/2015-04-28/libya-oilfield-write-off-by-total-is-ominous-sign-to-other-firms. Published April 28, 2015.

[76] Gosden, E. (2016). 'Shell Profits Drop on Shale Write-Down and Nigerian Woes'. *The Daily Telegraph*. http://www.telegraph.co.uk/finance/newsbysector/energy/oilandgas/10215535/Shell-profits-drop-on-shale-write-down-and-Nigerian-woes.html. Published September 15, 2016.

[77] Mufson, S. (2007). 'Conoco, Exxon Exit Venezuela Oil Deals'. *Washington Post*. http://www.washingtonpost.com/wp-dyn/content/article/2007/06/26/AR2007062602061.html. Published June 27, 2007.

[78] Mallet, R. (2015). 'The Long War: Venezuela and ExxonMobil'. *Telesur*. http://www.telesurtv.net/english/analysis/The-Long-War-Venezuela-and-ExxonMobil-20151118-0013.html. Published November 18, 2015.

[79] Connor, S. (2009). 'Warning: Oil Supplies Are Running Out Fast'. *The Independent*. http://www.independent.co.uk/news/science/warning-oil-supplies-are-running-out-fast-1766585.html. Published August 3, 2009.

[80] U.S. Department of the Treasury Office of Foreign Asset Contro. (2016). *Executive Order 13224 Blocking Terrorist Property and a Summary of the Terrorism Sanctions Regulations (Title 31 Part 595 of the U.S . Code of Federal Regulations), Terrorism List Governments Sanctions Regulations (Title 31 Part 596 of the U.S. Co.)*

[81] Mills, R. (2016). *Under the Mountains: Kurdish Oil and Regional Politics*. Published January, 2016.

[82] Rasheed, A. (2016). 'Update 2-Iraq Oil Min Says 2016 Development Costs Cut to 9 Billion USD'. *Reuters*. http://af.reuters.

com/article/commoditiesNews/idAFL8N1612Y8. Published
February 22, 2016.

[83] Chruickshank, P., N. Robertson. (2014). 'ISIS Comes to Libya'.
CNN. http://edition.cnn.com/2014/11/18/world/isis-libya/.
Published November 18, 2014.

[84] Nichols, M. (2016). 'ISIS in Libya "Could Relocate" from Sirte'.
Al Arabiya. http://english.alarabiya.net/en/News/middle-
east/2016/07/19/ISIS-in-Libya-could-relocate-from-Sirte-UN-
warns-.html. Published July 19, 2016.

[85] Graham-Harrison, E,, C. Stephen. (2016). 'Libyan Forces
Claim Sirte Port Captured From Isis As Street Battles Rage'.
The Guardian. https://www.theguardian.com/world/2016/
jun/10/libyan-forces-fight-street-battles-with-isis-for-control-
of-sirte. Published June 11, 2016.

[86] Brice, W. (2015). *Challenges with the Production of Heavier
Crudes: PDVSA Perspectives.*

[87] Editors, T. (2016). 'IMF Sees Venezuelan Inflation at
720% This Year'. *Financial Times*. http://www.ft.com/fast-
ft/2016/01/22/imf-sees-venezuela-inflation-at-720-this-year/.
Accessed May 16, 2016.

[88] Editors, T. (2015). 'Debt of Venezuelan State Oil Company
PdVSA Up 6.38%'. *El Universal*. http://www.eluniversal.com/
economia/150127/debt-of-venezuelan-state-oil-company-
pdvsa-up-638. Accessed May 15, 2016.

[89] *Venezuela and PDVSA Debt: A Guide.* (2016). Published
March, 2016.

[90] Baqi, M.M.A., N.G. Saleri. (2004). *Fifty-Year Crude Oil Supply
Scenarios: Saudi Aramco's Perspective.*

[91] Gamal, R.E. (2016). 'Saudi Aramco Finalizes IPO Options and
Plans for Global Expansion'. *Reuters*. http://www.reuters.
com/article/us-saudi-aramco-idUSKCN0Y10XL. Published
May 12, 2016.

[92] Editorial Staff. (2003). 'Saudi Aramco Spends to Stand Still'.
Petroleum Intelligence Weekly. Published December 15, 2003.

[93] Obaid, N. (2006). *Saudi Arabia's Strategic Energy Initiative:
Safeguarding Against Supply Disruptions*. https://csis-prod.
s3.amazonaws.com/s3fs-public/legacy_files/files/attach-
ments/061109_omsg_presentation_1.pdf.

[94] Husain, S.R. (2008). 'Kingdom Stands Vindicated After IEA Report on Ghawar'. *Arab News*. http://www.arabnews.com/node/318134. Published November 14, 2008.

[95] Henni, A. (2014). 'Saudi Aramco Wants Fields Fully Smart by 2017'. *Society of Petroleum Engineers*. http://www.spe.org/news/article/saudi-aramco-wants-fields-fully-smart-by-2017. Published June 27, 2014.

[96] Energy Information Administration. (2007). *International Energy Outlook*.

[97] IEA. (2007). *World Energy Outlook 2007*.

[98] Strahan, D. (2007). 'Oil Has Peaked, Prices to Soar'. *Last Oil Shock*. http://www.davidstrahan.com/blog/?p=67. Published October 29, 2007.

[99] U.S. Energy Information Administration. (1993). *Drilling Sideways: A Review of Horizontal Well Technology and Its Domestic Application*. Published April 30, 1993.

[100] York, S. (2016). 'Rebalancing of the Oil Market Begins'. Published March, 2016.

[101] Atkins, L., T. Higgins, C. Barnes. (2010). 'Heavy Crude Oil: A Global Analysis and Outlook to 2030'. Published November 12, 2010.

References Chapter 5

[1] Beament, E. (2016). 'World Felt Its Hottest May on Record, U.S. Scientists Reveal'. *The Press Association Limited*. Published June 16, 2016.

[2] NASA Goddard Institute for Space Studies. (2016). 'GISS Surface Temperature Analysis (GISTEMP)'. *NASA Website*. http://data.giss.nasa.gov/gistemp/. Published July 19, 2016.

[3] Hartmann, D.J., A.M.G. Klein Tank, M. Rusticucci, et. al. (2013). 'Observations: Atmosphere and Surface'. *Clim Chang 2013 Phys Sci Basis Contrib Work Gr I to Fifth Assess Rep Intergov Panel Clim Chang*: 159-254. DOI:10.1017/CBO9781107415324.008.

[4] NASA. (2016) *NASA Global Climate Change*. http://climate.nasa.gov/vital-signs/sea-level/.

[5]   Martín-Español, A., A. Zammit-Mangion, P.J. Clarke, et. al. (2016). 'Spatial and Temporal Antarctic Ice Sheet Mass Trends, Glacio-Isostatic Adjustment, and Surface Processes from a Joint Inversion of Satellite Altimeter, Gravity, and GPS Data'. *J Geophys Res F Earth Surf.* 2016;121(2):182-200. DOI:10.1002/2015JF003550.

[6]   Kjeldsen, K.K., N.J. Korsgaard, A.A. Bjørk, et. al. (2015). Spatial and Temporal Distribution of Mass Loss from the Greenland Ice Sheet Since AD 1900'. *Nature.* 2015;528(7582):396-400. DOI:10.1038/nature16183.

[7]   Enderlin, E., I. Howat. (2014). 'An Improved Mass Budget for the Greenland Ice Sheet'. *Geophys:*1-7. DOI:10.1002/2013GL059010.

[8]   Shepherd, A., E.R. Ivins, A.G. et. al. (2012). 'A Reconciled Estimate of Ice-Sheet Mass Balance'. *Science* 80. 2012;338(6111):1183-1189. DOI:10.1126/science.1228102.

[9]   Coumou, D., S. Rahmstorf. (2012). 'A Decade of Weather Extremes'. *Nat Clim Chang.* 2012;2(7):491-496. DOI:10.1038/Nclimate1452.

[10] Munich, R.E. (2016). 'Natural Loss Events Worldwide 2015'. *Geographical Overview.* Published January, 2016. https://www.munichre.com/site/wrap/get/documents_E1656163460/mram/assetpool.munichreamerica.wrap/PDF/07Press/2015_World_map_of_nat_cats.pdf.

[11] NOAA. (2015). 'How Many Tropical Cyclones Have There Been Each Year in the Atlantic Basin? What Years Were the Greatest and Fewest'. seen.http://www.aoml.noaa.gov/hrd/tcfaq/E11.html. Published June 1, 2015.

[12] Anagnostou, E., E.H. John, K.M. Edgar, et. al. (2016). 'Changing Atmospheric CO_2 Concentration Was the Primary Driver of Early Cenozoic Climate'. *Nature.* 2016;533(7603):1-19. DOI:10.1038/nature17423.

[13] Beerling, D.J., D.L. Royer. (2011). 'Convergent Cenozoic CO_2 History'. *Nat Geosci.* 2011;4(7):418-420. DOI:10.1038/ngeo1186.

[14] Etheridge, D.M. (2006). 'Law Dome Ice Core 2000-Year CO_2, CH_4, and N_2O Data'. *World Data Center for Paleoclimatology.* ftp://ftp.ncdc.noaa.gov/pub/data/paleo/icecore/antarctica/law/law2006.txt. Published 2006.

[15] Lisiecki, L.E., M.E. Raymo. (2005). 'A Pliocene-Pleistocene Stack of 57 Globally Distributed Benthic?'. *Paleoceanography*. 2005;20(1):1-17. DOI:10.1029/2004PA001071.

[16] Luthi, D., M. Le Floch, B. Bereiter, T. Blunier, U. Siegenthaler, D. Raynaud, J. Jouzel, H. Fischer, K.K. Stocker. (2008). 'EPICA Dome C Ice Core 800KYr Carbon Dioxide Data'. *World Data Center for Paleoclimatology*. http://www1.ncdc.noaa.gov/pub/data/paleo/icecore/antarctica/epica_domec/edc-co2-2008.txt.

[17] Monnin, E., E.J. Steig, U. Siegenthaler, K. Kawamura, J. Schwander BS, T.F. Stocker, D.L. Morse, J.M. Barnola, B. Bellier, D. Raynaud and HF. (2004). 'EPICA Dome C Ice Core High Resolution Holocene and Transition CO2 Data'. *World Data Center for Paleoclimatology*. ftp://ftp.ncdc.noaa.gov/pub/data/paleo/icecore/antarctica/epica_domec/edc-co2.txt.

[18] Royer, D.L. (2013). 'Atmospheric CO2 and O2 During the Phanerozoic: Tools, Patterns, and Impacts'. *Elsevier* 6. (2nd ed.). DOI:10.1016/B978-0-08-095975-7.01311-5.

[19] Schneider, R., J. Schmitt, P. Köhler, F. Joos, H. Fischer. (2013). 'A Reconstruction of Atmospheric Carbon Dioxide and its Stable Carbon Isotopic Composition from the Penultimate Glacial Maximum to the Last Glacial Inception'. *Clim Past*. 2013;9(6):2507-2523. DOI:10.5194/cp-9-2507-2013.

[20] Tans, P. (2016). 'Mauna Loa Carbon Dioxide'. *NOAA*. ftp://aftp.cmdl.noaa.gov/products/trends/co2/co2_annmean_mlo.txt.

[21] Zachos, J.C., G.R. Dickens, R.E. Zeebe. (2008). 'An Early Cenozoic Perspective on Greenhouse Warming and Carbon-Cycle Dynamics'. *Nature*. 2008;451(January):279-283. DOI:10.1038/nature06588.

[22] Zhang, Y.G., M. Pagani, Z. Liu, S.M. Bohaty, R. Deconto. (2013). 'A 40-Million-Year History of Atmospheric CO 2'. *Philos Trans R Soc*.

[23] Record, T.P., G.S. Change. (2005). 'The Phanerozoic Record of Global Sea-Level Change'. *Science* 80. 2005;310(5752):1293-1298. DOI:10.1126/science.1116412.

[24] Marcott, S.A., J.D. Shakun, P.U. Clark, A.C. Mix. (2013). 'A Reconstruction of Regional and Global Temperature for

the Past 11,300 Years'. *Science* 80. 2013;339(6124):1198-1201. DOI:10.1126/science.1228026.

[25] Ferguson, W. (2013). 'Ice Core Data Help Solve a Global Warming Mystery'. *Scientific American*. http://www.scientificamerican.com/article/ice-core-data-help-solve/. Published March 1, 2013.

[26] Shakun, J.D., P.U. Clark, F. He, et. al. (2012). 'Global Warming Preceded by Increasing Carbon Dioxide Concentrations During the Last Deglaciation'. *Nature*. 2012;484(7392):49-54. DOI:10.1038/nature10915.

[27] del Giorgio PA, Duarte CM. (2002). Respiration in the open ocean. *Nature*. 2002;420(6914):379-384. DOI:10.1038/nature01165.

[28] Tripati, A.K., C.D. Roberts, R. Eagle. (2009). 'Coupling of CO_2 and Ice Sheet Stability over Major Climate Transitions of the Last 20 Million Years'. *Science*. 2009;326(5958):1394-1397. DOI:10.1126/science.1178296.

[29] Pross, J., L. Contreras, P.K. Bijl, et. al. (2012). 'Persistent Near-Tropical Warmth on the Antarctic Continent During the Early Eocene Epoch'. *Nature*. 2012;488(7409):73-77. DOI:10.1038/nature11300.

[30] Pagani, M., K. Caldeira, R. Berner, D.J. Beerling. (2009). 'The Role of Terrestrial Plants in Limiting Atmospheric CO_2 Decline Over the Past 24 Million Years'. *Nature*. 2009;460(7251):85-88. DOI:10.1038/nature08133.

[31] Jagniecki, E.A., T.K.T.K. Lowenstein, D.M. Jenkins, R.V. Demicco. (2015). 'Eocene Atmospheric CO_2 from the Nahcolite Proxy'. *Geology*. 2015;43(12):1075-1078. DOI:10.1130/G36886.1.

[32] Reviewed, P., D, Jeffrey, J. Timothy, A. Philip. (2009). 'Previously Published Works'. *UC Riverside*. 2009;(2014). DOI:10.1016/j.cognition.2008.05.007.

[33] Pagani, M., J.C. Zachos, K.H. Freeman, B. Tripple, S. Bohaty. (2005). 'Marked Decline in Atmospheric Carbon Dioxide Concentrations During the Paleocene'. *Science* 80. 2005;309(5734):600-603. DOI:10.1126/science.1110063.

[34] Kahn, B. (2016). 'Antarctic CO_2 Hit 400 PPM for First Time in 4 Million Years'. *Scientific American*. http://www.scientifi-

camerican.com/article/antarctic-co2-hit-400-ppm-for-first-time-in-4-million-years/. Published June 16, 2016.

[35] Ganopolski, A., R. Winkelmann, H.J. Schellnhuber. (2015). 'Critical Insolation-CO2 Relation for Diagnosing Past and Future Glacial Inception'. *Begutachtung.* 2015;529(7585):200-203. DOI:10.1038/nature16494.

[36] Inman, M. (2008). 'Carbon is Forever'. *Nature Reports Climate Change.* http://www.nature.com/climate/2008/0812/full/climate.2008.122.html. Published November 20, 2008.

[37] Archer, D., M. Eby, V. Brovkin, et. al. (2009). 'Atmospheric Lifetime of Fossil Fuel Carbon Dioxide'. *Annu Rev Earth Planet Sci.* 2009;37(1):117-134. DOI:10.1146/annurev.earth.031208.100206.

[38] Grimes, S. (2015). 'Palaeoclimate: Carbon Feedbacks on Repeat?'. *Nat Geosci.* 2015;8(1):7-8. DOI:10.1038/ngeo2337.

[39] Bowen, G.J., J.C. Zachos. (2010). 'Rapid Carbon Sequestration at the Termination of the Palaeocene–Eocene Thermal Maximum'. *Nat Geosci.* 2010;3(12):866-869. DOI:10.1038/ngeo1014.

[40] Cui, Y., L.R. Kump, A.J. Ridgwell, et. al. (2011). 'Slow Release of Fossil Carbon During the Palaeocene–Eocene Thermal Maximum'. *Nat Geosci.* 2011;4(7):481-485. DOI:10.1038/ngeo1179.

[41] Bowen, G.J., B.J. Maibauer, M.J. Kraus, et. al. (2015). 'Two Massive, Rapid Releases of Carbon During the Onset of the Palaeocene-Eocene Thermal Maximum'. *Nat Geosci.* 2015;8(1):44-47. DOI:10.1038/ngeo2316.

[42] Zeebe, R.E., A. Ridgwell, J.C. Zachos. (2016). 'Anthropogenic Carbon Release Rate Unprecedented During the Past 66 Million Years'. *Nat Geosci.* 2016;9(April):325-329. DOI:10.1038/ngeo2681.

[43] Intergovernmental Panel on Climate Change. (2013). IPCC Fifth Assessment Report. https://www.ipcc.ch/report/ar5/.

[44] Urban, T. (2015). 'How Tesla Will Change the World'. *Wait But Why.* http://waitbutwhy.com/2015/06/how-tesla-will-change-your-life.html. Published June 2, 2015.

[45] Church, J., P.U. Clark, A. Cazenave, et. al. (2013). 'Sea Level Change'. *Clim Chang 2013 Phys Sci Basis Contrib Work Gr I to Fifth Assess Rep Intergov Panel Clim Chang*:1137-1216. DOI:10.1017/CBO9781107415315.026.

[46] Munich, R.E. (2016). 'Climate Insurance: A Stepping Stone to Sustainable Growth'. https://www.munichre.com/en/reinsurance/magazine/topics-online/2016/topicsgeo2015/climate-insurance/index.html?QUERYSTRING=*. Published March 2, 2016.

[47] Hoeppe, P. (2016). 'Trends in Weather Related Disasters: Consequences for Insurers and Society. *Weather Clim Extrem*. 2016;11:70-79. DOI:10.1016/j.wace.2015.10.002.

[48] *Answers and Demands of German Insurers.*

[49] Al Jazeera. (2016) 'Middle East in the Grip of a Major Heatwave'. http://www.aljazeera.com/news/2016/07/middle-east-grips-major-heatwave-160723083445674.html. Published July 23, 2016 .

[50] Lelieveld, J., Y. Proestos, P. Hadjinicolaou, M. Tanarhte, E. Tyrlis, G. Zittis. (2016). 'Strongly Increasing Heat Extremes in the Middle East and North Africa (MENA) in the 21st Century'. *Clim Change*:1-16. DOI:10.1007/s10584-016-1665-6.

[51] Pal, J.S., E.A.B. Eltahir. (2015). 'Future Temperature in Southwest Asia Projected to Exceed a Threshold for Human Adaptability'. *Nat Clim Chang*. 2015;18203(October):1-4. DOI:10.1038/nclimate2833.

[52] Bellard, C., C. Leclerc, F. Courchamp. (2014). 'Impact of Sea Level Rise on the 10 Insular Biodiversity Hotspots'. *Glob Ecol Biogeogr*. 2014;23(2):203-212. DOI:10.1111/geb.12093.

[53] Kench, P.S., D. Thompson, M.R. Ford, H. Ogawa, R.L. McLean. (2015). 'Coral Islands Defy Sea-Level Rise over the Past Century: Records from a Central Pacific Atoll'. *Geology*. 2015;43(6):515-518. DOI:10.1130/G36555.1.

[54] Nicholls, R.J., F.M.J. Hoozemans, M. Marchand. (1999). 'Increasing Flood Risk and Wetland Losses Due to Global Sea-Level Rise: Regional and Global Analyses'. *Glob Environ Chang*. 1999;9(SUPPL.). DOI:10.1016/S0959-3780(99)00019-9.

[55] Nicholls, R.J., S. Hanson, C. Herweijer, et. al. (2007). 'Ranking Port Cities with High Exposure and Vulnerability to Climate Extremes: Exposure Estimates'. *Environment*. 2007;1(1):53-57. DOI:10.1787/011766488208.

[56] Lukyanets, A.S., T.N. Khanh, S.V. Ryazantsev, V.S. Tikunov, P. Hoang Hai. (2015). 'Influence of Climatic Changes on Popula-

tion Migration in Vietnam'. *Geogr Nat Resour*. 2015;3(3):191-196. DOI:10.1134/S1875372815030129.

[57] Roberts, C. (2013). *Ocean of Life*. London: Penguin.

[58] Fabry, V.J., J.C. Orr, O. Aumont, et. al. (2005). 'Anthropogenic Ocean Acidification over the Twenty-First Century and its Impact on Calcifying Organisms'. *Nature*. 2005;437(7059):681-686. DOI:10.1038/nature04095.

[59] Harvey, B.P., D. Gwynn-Jones, P.J. Moore. (2013). 'Meta-Analysis Reveals Complex Marine Biological Responses to the Interactive Effects of Ocean Acidification and Warming'. *Ecol Evol*. 2013;3(4):1016-1030. DOI:10.1002/ece3.516.

[60] Bopp, L., L. Resplandy, J.C. Orr, et. al. (2013). 'Multiple Stressors of Ocean Ecosystems in the 21st Century: Projections with CMIP5 Models'. *Biogeosciences*. 2013;10(10):6225-6245. DOI:10.5194/bg-10-6225-2013.

[61] Branch, T.A., B.M. DeJoseph, L.J. Ray, C.A. Wagner. (2013). 'Impacts of Ocean Acidification on Marine Seafood'. *Trends Ecol Evol*. 2013;28(3):178-186. DOI:10.1016/j.tree.2012.10.001.

[62] Kroeker, K.J., R.L. Kordas, R. Crim, et. al. (2013). 'Impacts of Ocean Acidification on Marine Organisms: Quantifying Sensitivities and Interaction with Warming'. *Glob Chang Biol*. 2013;19(6):1884-1896. DOI:10.1111/gcb.12179.

[63] Maier, C., A. Schubert, S. Berzunza, M.M. Sanchez, M.G. Weinbauer, P. Watremez, J.P. Gattuso. (2013). 'End of the Century pCO_2 Levels Do Not Impact Calcification in Mediterranean Cold-Water Corals'. *PLoS One*. 2013;8(4). DOI:10.1371/journal.pone.0062655.

[64] Slezak, M. (2016). 'The Great Barrier Reef: A Catastrophe Laid Bare'. *The Guardian*. https://www.theguardian.com/environment/2016/jun/07/the-great-barrier-reef-a-catastrophe-laid-bare. Published June 7, 2016.

[65] Hooidonk, R. van, J.A. Maynard, S. Planes. (2013). 'Temporary Refugia for Coral Reefs in a Warming World'. *Nat Clim Chang*. 2013;3(5):508-511. DOI:10.1038/nclimate1829.

[66] Kimball, B.A. (2016). 'Crop Responses to Elevated CO_2 and Interactions with H_2O, N, and Temperature'. *Curr Opin Plant Biol*. 2016;31:36-43. DOI:10.1016/j.pbi.2016.03.006.

[67] Knox, J., T. Hess, A. Daccache, T. Wheeler. (2012). 'Climate Change Impacts on Crop Productivity in Africa and South Asia'. *Environ Res Lett*. 2012;7(3):34032. DOI:10.1088/1748-9326/7/3/034032.

[68] Deryng, D., D. Conway, N. Ramankutty, J. Price, R. Warren. (2014). 'Global Crop Yield Response to Extreme Heat Stress Under Multiple Climate Change Futures'. *Environ Res Lett Environ Res Lett*. 2014;9:34011-34013. DOI:10.1088/1748-9326/9/3/034011.

[69] Rosenzweig, C., J. Elliott, D. Deryng, et. al. (2014). 'Assessing Agricultural Risks of Climate Change in the 21st Century in a Global Gridded Crop Model Intercomparison'. *Proc Natl Acad Sci U S A*. 2014;111(9):3268-3273. DOI:10.1073/pnas.1222463110.

[70] Cook, J., N. Oreskes, P.T. Doran, et. al. (2016). 'Consensus on Consensus: A Synthesis of Consensus Estimates on Human – Caused Global Warming'. *Environ Res Lett*. 2016;11(2016):1-24. DOI:10.1088/1748-9326/11/4/048002.

[71] Tol, R.S.J. (2014). 'Quantifying the Consensus on Anthropogenic Global Warming in the Literature: Rejoinder'. *Energy Policy*. 2014;73(4):709. DOI:10.1016/j.enpol.2014.06.003.

[72] Haigh, J., P. Cargill. (2015). *The Sun's Influence on Climate*. Princeton University Press.

[73] Haigh, J.D. (2002). 'The Effects of Solar Variability on the Earth's Climate'. *R Soc*. 2002;361(1802):95-111. DOI:10.1098/rsta.2002.1111.

[74] Wang, K., R.E. Dickinson. (2013). 'Contribution of Solar Radiation to Decadal Temperature Variability over Land'. *Proc Natl Acad Sci U S A*. 2013;110(37):14877-14882. DOI:10.1073/pnas.1311433110.

[75] Haigh, J. (2011). 'Solar Influences on Climate'. *Grantham Brief Pap*. 2011;(5):1-20.

[76] Kroviva, N.A. (2015). *Daily Reconstruction of Solar Irradiance Since 1960*. http://www2.mps.mpg.de/projects/sun-climate/data/SATIRE-T_SATIRE-S_TSI_20150406.txt.

[77] Andrews, T., J.M. Gregory, M.J. Webb, K.E. Taylor. (2012). 'Forcing, Feedbacks and Climate Sensitivity in CMIP5 Coupled Atmosphere-Ocean Climate Models'. *Geophys Res Lett*. 2012;39(9):1-7. DOI:10.1029/2012GL051607.

[78] Marvel, K., G.A. Schmidt, R.L. Miller, L.S. Nazarenko. (2015). 'Implications for Climate Sensitivity from the Response to Individual Forcings'. *Nat Clim Chang.* 2015;6(April):3-6. DOI:10.1038/nclimate2888.

[79] Richardson, M., K. Cowtan, E. Hawkins, M.B. Stolpe. (2016). 'Reconciled Climate Response Estimates from Climate Models and the Energy Budget of Earth'. *Nat Clim Chang.* 2016;(June):1-6. DOI:10.1038/nclimate3066.

[80] Gillett, N.P., V.K. Arora, D. Matthews, M.R. Allen. (2013). 'Constraining the Ratio of Global Warming to Cumulative CO_2 Emissions Using CMIP5 Simulations'. *J Clim.* 2013;26(18):6844-6858. DOI:10.1175/JCLI-D-12-00476.1.

[81] MacDougall, A.H. (2015). 'The Transient Response to Cumulative CO_2 Emissions: a Review'. *Curr Clim Chang Reports.* 2015:39-47. DOI:10.1007/s40641-015-0030-6.

[82] Xu, J.H., Y. Fan, S.M. Yu. (2014). 'Energy Conservation and CO_2 Emission Reduction in China's 11th Five-Year Plan: A Performance Evaluation'. *Energy Econ.* 2014;46:348-359. DOI:10.1016/j.eneco.2014.10.013.

[83] D'aprile, A., M. Davide. (2016). *China Policy Highlights: China's 13th Five-Year Plan and Other Recent Developments*: 24.

[84] Qi, Y., N. Stern, T. Wu, J. Lu, F. Green. (2016). 'China's Post-Coal Growth'. *Nat Geosci.* DOI:10.1038/ngeo2777.

[85] U.S. Environmental Protection Agency. (2015). *Carbon Pollution Emission Guidelines for Existing Stationary Sources: Electric Utility Generating Units* 80.

[86] McGlade, C., P. Ekins. (2015). 'The Geographical Distribution of Fossil Fuels Unused When Limiting Global Warming to $2°C$'. *Nature.* 2015;517(7533):187-190. DOI:10.1038/nature14016.

[87] McGlade, C., P. Ekins. (2014). 'Un-Burnable Oil: An Examination of Oil Resource Utilisation in a Decarbonised Energy System'. *Energy Policy.* 2014;64:102-112. DOI:10.1016/j.enpol.2013.09.042.

[88] Rystad Energy. (2013). Petroleum Production Under the Two Degree Scenario (2DS)':1-39.

[89] Hope, M. (2015). *Carbon Supply Cost Surves Oil.* http://www.energypost.eu/implications-50-barrel-oil-worlds-energy-

mix/\npapers3://publication/uuid/DE7D6771-F4EC-479D-978F-3DE743790454.

[90] Carbon Tracker Initiative. (2015). *The $2 Trillion Stranded Assets Danger Zone: How Fossil Fuel Firms Risk Destroying Investor Returns.* http://www.carbontracker.org/wp-content/uploads/2015/11/CAR3817_Synthesis_Report_24.11.15_WEB2.pdf.

[91] UNEP Finance Initiative, CDP. (2016). *From Disclosure to Action: The First Annual Report of the Portfolio Decarbonization Coalition.*

[92] Allianz, S.E. (2015). 'Climate Protection will Become Part of Core Business'. *Webpage.* https://www.allianz.com/en/press/news/financials/stakes_investments/151126_climate-protection-will-become-part-of-core-business/. Published November 26, 2015.

[93] Carrington, D. (2015). 'Norway Confirms 900 Billion Sovereign Wealth Fund's Major Coal Divestment'. *The Guardian.* https://www.theguardian.com/environment/2015/jun/05/norways-pension-fund-to-divest-8bn-from-coal-a-new-analysis-shows. Published June 5, 2015.

[94] Hansen, J., M. Sato, P. Hearty, et. al. (2016). 'Ice Melt, Sea Level Rise and Superstorms: Evidence from Paleoclimate Data, Climate Modeling, and Modern Observations that Global Warming Could be Dangerous'. *Atmos Chem Phys.* 2016;16(6):3761-3812. DOI:10.5194/acp-16-3761-2016.

[95] Jevrejeva, S., A. Grinsted, J.C. Moore. (2014). 'Upper Limit for Sea Level Projections by 2100'. *Environ Res Lett.* 2014;9(10):104008. DOI:10.1088/1748-9326/9/10/104008.

[96] Riyadh. (2015). The Intended Nationally Determined Contribution of the Kingdom of Saudi Arabia under the UNFCCC. 2015;(November):1-7. http://www4.unfccc.int/submissions/INDC/Published Documents/Saudi Arabia/1/KSA-INDCs English.pdf.

References Chapter 6

[1]   China Daily. (2015). Full text of President Xi's speech at opening ceremony of Paris climate summit. *China Daily.* http://www.chinadaily.com.cn/world/XiattendsParisclimateconference/2015-12/01/content_22592469.htm. Published January 12, 2015.

[2]   Johnson, J., A. Lefebvre. (2003). 'U.S.: Impact of Northeast Blackout Continues to Emerge'. *World Socialist Web Site.* https://www.wsws.org/en/articles/2003/08/blck-a20.html. Published August 20, 2003.

[3]   Liscouski, B., W. Elliot. (2004). *Final Report on the August 14, 2003 Blackout in the United States and Canada: Causes and Recommendations.* https://reports.energy.gov/BlackoutFinal-Web.pdf.

[4]   Bank, T.W. (2016). 'Infrastructure Reliability and Availability Dataset'. *World Bank Group Enterprise Surveys.* http://www.enterprisesurveys.org/data/exploretopics/infrastructure.

[5]   Loughborough University Institutional Repository. (2010). 'Domestic Electricity Use: A High Resolution Energy Demand Model'. 2010;42(10):1878-1887.

[6]   Department of Energy & Climate Change. (2016). *Energy Trends June 2016.*

[7]   BP. (2016). 'BP Statistical Review of World Energy June 2016'. http://www.bp.com/statisticalreview.

[8]   U.S. Energy Information Administration. (2016). 'Shale in the United States'. *Energy in Brief.* https://www.eia.gov/energy_in_brief/article/shale_in_the_united_states.cfm#shaledata.

[9]   U.S. Energy Information Administration. (2015). 'U.S. Crude Oil and Natural Gas Proved Reserves'. http://www.eia.gov/naturalgas/crudeoilreserves/. Published November 23, 2015.

[10]   Joshi, K. (2015). *Forecasting Production in Shale Gas Reservoirs: A Better Assessment of Reserves.*

[11]   U.S. Energy Information Administration. (2016). 'Natural Gas Weekly Update for Week Ending August 3, 2016'. *Natural Gas Weekly Update.* http://www.eia.gov/naturalgas/weekly/. Published August 4, 2016.

[12]   Kaminski, B. (1996). *Economic Transition in Russia and the New States of Eurasia*. New York: M.E. Sharpe.

[13]   Russian Federation Federal State Statistics Service. (2016). 'Russia Producer Price Indices by Economic Activity'. http://www.gks.ru/wps/wcm/connect/rosstat_main/rosstat/en/figures/prices/.

[14]   Stern, D. (2013). 'The Establishment of Natural Gas Prices: Past, Present and Future'. *The HSE Economic Journal*. http://library.hse.ru/e-resources/HSE_economic_journal/articles/17_03_06.pdf.

[15]   TrendEconomy. (2016). 'The Price of Natural Gas from Russia'. http://data.trendeconomy.ru/imf/weo/Natural_Gas_Russian_Natural_Gas_border_price_in_Germany_PNGASEU/World?imf_weo_version=20160412. Published December 4, 2016.

[16]   U.S. Energy Information Administration. (2016). 'U.S. Natural Gas Wellhead Price'. https://www.eia.gov/dnav/ng/hist/n9190us3m.htm.

[17]   U.S. Energy Information Administration. (2016). 'Natural Gas Prices'. https://www.eia.gov/dnav/ng/ng_pri_sum_dcu_nus_m.htm.

[18]   Wood Mackenzie. (2014). 'LNG AT50'. *InfoGraphic*. https://www.woodmac.com/content/portal/energy/highlights/wk2_Oct_14/LNGat50 InfographicDownload.pdf.

[19]   Statistik der Kohlenwirtschaft e.v. (2016). 'Entwicklung Ausgewählter Energiepreise'. *Webpage*. http://www.kohlenstatistik.de/17-0-Deutschland.html. Published October, 2016.

[20]   Inman, M. (2016). 'Can Fracking Power Europe?'. *Nature*. 2016;531(7592):22-24. DOI:10.1038/531022a.

[21]   'Russia Won't Develop Shale Gas for a Decade'. (2013). *The Petroleum Economist*. http://www.petroleum-economist.com/articles/politics-economics/asia-pacific/2013/russia-wont-develop-shale-gas-for-a-decade. Published April 19, 2013.

[22]   Guo, A. (2015). 'PetroChina, Sinopec Shale Gas Output Said to be Below China's Goal'. *World Oil*. http://www.worldoil.com/news/2015/12/09/petrochina-sinopec-shale-gas-output-said-to-be-below-china-s-goal. Published September 12, 2015.

[23] 'Sinopec's Fuling Shale Gas Field Production in H1 Sees Sharp Jump'. (2016). *Natural Gas Asia*. Published July 7, 2016.

[24] Deborah, G., Y. Sautin, W. Tao. (2014). *China's Oil Future*. http://carnegieendowment.org/2014/05/06/china-s-oil-future.

[25] Hu, A., Q. Dong. (2015). 'On Natural Gas Pricing Reform in China'. *Nat Gas Ind*. 2015;35(4):99-106. DOI:10.3787/j.issn.1000-0976.2015.04.016.

[26] Ambrose, J. (2016). 'BP Deepens Chinese Shale Gas Interests'. *The Daily Telegraph*. http://www.telegraph.co.uk/business/2016/09/01/bp-deepens-chinese-shale-gas-interests/. Published September 1, 2016.

[27] Kuukskraa, V.A., T.E. Hoak, J.A. Kuuskraa. (1996). 'Exploration Tight Sands Gain as U.S. Gas Source'. *Oil Gas J*. Published March, 1996. http://www.ogj.com/articles/print/volume-94/issue-12/in-this-issue/exploration/exploration-tight-sands-gain-as-us-gas-source.html.

[28] Administration EI. (2016). *Annual Energy Outlook 2016: Energy Production, Imports and Exports*. https://www.eia.gov/forecasts/aeo/section_energyprod.cfm.

[29] Agency IE. (2016). 'IEA Unconventional Gas Production Database'. *Database*. http://www.iea.org/ugforum/ugd/.

[30] Dong, Z., S.A. Holditch, W.J. Lee. (2016). 'World Recoverable Unconventional Gas Resource Assessment'. *Unconventional Oil and Gas Resources Handbook*. (1st ed.) *Elsevier*; 2016:53-70.

[31] United States Geological Survey. (2012). 'An Estimate of Undiscovered Conventional Oil and Gas Resources'. *USGS Fact Sheet*. 2012;2012–3042(March):6. http://pubs.usgs.gov/fs/2012/3042/.

[32] Chew, K.J. (2013). *The Future of Oil: Unconventional Fossil Fuels Subject Areas*.

[33] Report, S.I. (2015). *U.S. Geological Survey Assessment of Reserve Growth Outside of the United States*.

[34] WEC. (2013). 'World Energy Resources'. *World Energy Counc Rep*.:468. DOI:http://www.worldenergy.org/wp-content/uploads/2013/09/Complete_WER_2013_Survey.pdf.

[35] BGR. (2014). Energy Study 2014. *Reserv Resour Availab energy Resour*. 2014;18:- 131.

[36] TOTAL S.A. (2008). *The Know-How Series.*

[37] Summers, D. (2014). 'The Differences in Fracking Tight Sand and Shales'. *Oilprice.com.* http://oilprice.com/Energy/Natural-Gas/The-Differences-In-Fracking-Tight-Sand-And-Shales.html. Published August 6, 2014.

[38] 'What is Tight Gas, and How Is It Produced?'. *Rigzone.* http://www.rigzone.com/training/insight.asp?insight_id=346.

[39] Wang, H., R. Rezaee, A. Saeedi. (2016). 'Preliminary Study of Improving Reservoir Quality of Tight Gas Sands in the Near Wellbore Region by Microwave Heating'. *J Nat Gas Sci Eng.* 2016;32:395-406. DOI:10.1016/j.jngse.2016.04.041.

[40] Kelly, E. (2015). *Production Outlook by Major Basin , and Facilities Options, Edward Kelly, IHS Inc.*

[41] Xi, S., X. Liu, P. Meng. (2015). 'Exploration Practices and Prospect of Upper Paleozoic Giant Gas Fields in the Ordos Basin'. *Nat Gas Ind B.* 2015;2(5):430-439. DOI:10.1016/j.ngib.2015.09.019.

[42] Wang, J., S. Mohr, L. Feng, H. Liu, G.E. Tverberg. (2016). 'Analysis of Resource Potential for China's Unconventional Gas and Forecast for its Long-Term Production Growth'. *Energy Policy.* 2016;88:389-401. DOI:10.1016/j.enpol.2015.10.042.

[43] U.S. Energy Information Administration. (2016). *Annual Energy Outlook 2016 Early Release: Annotated Summary of Two Cases.*

[44] National Energy Board of Canada. (2016). *Canada' S Energy Future 2016.*

[45] China National Energy Board. *National Energy Board Released the Total Electricity Consumption in 2010-2015.*

[46] China Electricity Council. (2014). 'Thermal Power Installed Capacity Has Dropped Steadily'. *Webpage.* http://www.cec.org.cn/xinwenpingxi/2014-01-21/115810.html. Published January 21, 2014.

[47] IEA. (2011). *Are We Entering a Golden Age of Gas?* DOI:10.1049/ep.1977.0180.

[48] Li, X. (2015). *Natural Gas in China : A Regional Analysis.*

[49] Guo, A. (2016). 'Sinopec to Double Natural Gas Output by 2020 as China Shuns Coal'. *Bloomberg.* http://www.bloomberg.com/news/articles/2016-03-30/sinopec-to-double-

natural-gas-output-by-2020-as-china-shuns-coal. Published March 30, 2016.

[50] China to Raise Gas Imports by 500%. *RT*. https://www.rt.com/business/354615-china-gas-import-2030/.

[51] Platts. (2017). 'China Targets 13% Rise in Natural Gas, 7% Fall in Crude Output for 2016'. *Platts*. http://www.platts.com/latest-news/natural-gas/singapore/china-targets-13-rise-in-natural-gas-7-fall-in-26411759. Published April 6, 2017.

[52] Rose, A. (2016). 'China's 2015 Natural Gas Output Growth Slowest in at Least 10 Years'. *Reuters*. http://www.reuters.com/article/china-economy-output-gas-idUSL3N1532HZ. Published January 19, 2016.

[53] Aden, N., D. Fridley, N. Zheng. (2009). 'China's Coal: Demand, Constraints, and Externalities'. *Lawrence Berkeley Natl Lab*. Published July, 2009. http://escholarship.org/uc/item/8kk9n3rr.

[54] Koppelaar, R. (2009). 'Are Reserves of the Largest US Coal Field Overstated by 50%'. *The Oil Drum*. http://www.theoildrum.com/node/5122. Published February 24, 2009.

[55] IEA. (2015). *CO2 Emissions from Fossil Fuels: Highlights*. DOI:10.1787/co2-table-2011-1-en.

[56] EIA. (2016). 'Germany's Renewables Electricity Generation Grows in 2015, but Coal Still Dominant'. http://www.eia.gov/todayinenergy/detail.cfm?id=26372.

[57] IEA. (2016). *World Energy Investment 2016*. https://www.iea.org/investment/.

[58] Astakhova, O., C. Aizhu. (2016). 'Exclusive: Russia Likely to Scale Down China Gas Supply Plans'. *Reuters*. http://www.reuters.com/article/us-russia-china-gas-exclusive-idUSKC-N0UT1LG. Published January 15, 2016.

[59] IGU. (2016). *World LNG Report*: 88.

[60] Pradhan, D. (2016). 'Revised Qatar LNG Deal Cuts Gas Price to Below 5 USD per MMBTU'. *The Economic Times of India*. Published May 3, 2016.

[61] Union, I.G. (2015). *World LNG Report: 2015 Edition*: 100.

[62] Hsu, J.W. (2016). 'Asia's Liquified Natural Gas Prices Hit the Brakes'. *The Wall Street Journal*. http://www.wsj.com/articles/asias-liquefied-natural-gas-prices-hit-the-brakes-1461586616.

[63] Platts. (2011). *Coal Trader International* 11.

[64] COALSpot. (2014). 'Indonesian Coal Price Reference
      Crashes Through 65 USD'. *Webpage*. http://www.coalspot.
      com/news/3686/indonesian-coal-price-reference-crashes-
      through-65/0. Published December 15, 2014.

[65] Alstom. (2014). 'Management Report on Consolidated Finan-
      cial Statements-Fiscal Year 2014/15'.: 348.

[66] Xu, J.H., Y. Fan, S.M. Yu. (2014). 'Energy Conservation and
      CO2 Emission Reduction in China's 11th Five-Year Plan:
      A Performance Evaluation'. *Energy Econ*. 2014;46:348-359.
      DOI:10.1016/j.eneco.2014.10.013.

[67] D'aprile, A., M. Davide. (2016). *China Policy Highlights: Chi-
      na's 13th Five-Year Plan and Other Recent Developments*.: 24.

[68] Qi, Y., N. Stern, T. Wu, J. Lu, F. Green. (2016). 'China's Post-
      Coal Growth'. *Nat Geosci*. DOI:10.1038/ngeo2777.

[69] China National Energy Board. (2016). National Energy Board
      Released the Total Electricity Consumption in 2015. Pub-
      lished January 15, 2016.

[70] Ping, H. (2016). 'National Raw Coal Production Year on Year
      Decline in a Row in June over 10%'. *SXCoal*. http://www.
      sxcoal.com/news/4548391/info. Published October 24, 2016.

[71] 'China to Slow Energy Project Construction Shifts to Capac-
      ity Cuts'. (2016). *Reuters*. http://indianexpress.com/article/
      world/world-news/china-to-slow-energy-project-construc-
      tion-shifts-to-capacity-cuts-2910501/. Published July 13, 2016.

[72] Jaiswal, S. (2017). 'India Budget 2016: Renewables to Gain
      from Coal Tax Hike'. 2017;(March 2016):1-7.

[73] Goswami, U. (2016). 'India's Renewable Energy Targets
      Catch the Attention of Global Investors, Still Needs Ground
      Work'. *The Economic Times of India*. http://economictimes.
      indiatimes.com/news/politics-and-nation/indias-renewable-
      energy-targets-catch-the-attention-of-global-investors-still-
      need-ground-work/articleshow/53015707.cms. Published July
      2, 2016.

[74] Jai, S. (2016). 'Centre Scales Down Power Demand Forecast'.
      Business Standard. http://www.business-standard.com/
      article/economy-policy/centre-scales-down-power-demand-
      forecast-116042500050_1.html. Published April 25, 2016.

[75] Williams, D. (2016). 'India Cancels Plans for 16 GW of Coal Power'. *Power Engineering International.*

[76] Economic Times Bureau. (2016). 'India Won't Need Extra Power Plants for Next Three Years, Says Government Report'. *The Economic Times of India.* http://economictimes.indiatimes.com/industry/energy/power/india-wont-need-extra-power-plants-for-next-three-years-says-government-report/articleshow/52545715.cms?utm_source=contentofinterest&utm_medium=text&utm_campaign=cppst. Published June 2, 2016.

[77] Lowrey, D. (2015). 'Goldman Makes Case for "Peak Coal", Expects Pricing Pressure and Demand Output Fall'. *SNL.* https://www.snl.com/InteractiveX/Article.aspx?cdid=A-33970119-12844&mkt_tok=3RkMMJWWfF9wsRojvKjOce%2FhmjTEU5z17u8vUa%2B%2Bi4kz2EFye%2BLIHETpodcMSMJiMr3YDBceEJhqyQJxPr3FJNANysRuRhDgCw%3D%3D. Published September 23, 2015.

[78] World Nuclear. (2016). 'Nuclear Power in Japan'. *World Nucl Inf Libr.* June 2016. http://www.world-nuclear.org/information-library/country-profiles/countries-g-n/japan-nuclear-power.aspx.

[79] Editor. (2015). 'So Far, So Good? The French Energy Transition Law in the Starting Blocks'. *Energy Transition: The German Energiewende.* http://energytransition.de/2015/07/french-energy-transition-law/. Published July 29, 2015.

[80] World Nuclear Assocation. (2016). 'World Nuclear Power Reactors & Uranium Requirements'. *Webpage.* http://www.world-nuclear.org/information-library/facts-and-figures/world-nuclear-power-reactors-and-uranium-requireme.aspx. Published September 1, 2016.

[81] Schneider, M., A. Froggatt, J. Hazemann, T. Katsuta, S. Thomas. (2015). *The World Nuclear Industry-Status Report 2015.*:1-202.

[82] Larson, A. (2016). 'US Nuclear Power Plant Closures'. *Power.* http://www.powermag.com/u-s-nuclear-power-plant-closures-slideshow/. Published June 25, 2016.

[83] Steckler, N., D. Arnold, D. Besse, et. al. (2017). 'Reactors in the Red: Financial Health of the U.S. Nuclear Fleet by the Numbers'. 2017;(July 2016):1-16.

[84] Beaupuy, F. (2016). 'UK Approves EDF's 18 Billion Pounds Hinkley Point Nuclear Project'. *Bloomberg*. Published September 14, 2016.

[85] APX Power Exchange. (2016). 'UKPX RPD Historical Data'. *Webpage*. http://www.apxgroup.com/market-results/apx-power-uk/ukpx-rpd-historical-data/.

[86] Chetal, S.C., V. Balasubramaniyan, P. Chellapandi, et. al. (2006). 'The Design of the Prototype Fast Breeder Reactor'. *Nucl Eng Des*. 2006;236(7-8):852-860. DOI:10.1016/j.nucengdes.2005.09.025.

[87] Nathan, S. (2013). 'Prism Project: A Proposal for the UK's Problem Plutonium'. *The Engineer*. https://www.theengineer.co.uk/issues/energy-and-sustainability-special/prism-project-a-proposal-for-the-uks-problem-plutonium/. Published May 13, 2013.

[88] Wang, B. (2014). '800 MW Fast Neutron Russian Breeder Reactor is Fully Powered Up'. *Next Big Future*. http://nextbigfuture.com/2014/06/800-mw-fast-neutron-russian-breeder.html. Published June 27, 2014.

[89] Martin, R. (2016). 'China Could have a Meltdown-Proof Nuclear Reactor Next Year'. *MIT Technology Review*. https://www.technologyreview.com/s/600757/china-could-have-a-meltdown-proof-nuclear-reactor-next-year/. Published February 11, 2016.

[90] Butler, D. (2014). 'ITER's New Chief Will Shake Up Troubled Fusion Reactor'. *Nature*. http://www.nature.com/news/iter-s-new-chief-will-shake-up-troubled-fusion-reactor-1.16396. Published November 21, 2014.

[91] Clercq, G. de. (2016). '"Totally Unrealistic:" The International Nuclear Fusion Reactor Prototype Project is a Decade Late and 4 Billion Euro over Budget'. *Business Insider*. http://uk.businessinsider.com/r-iter-nuclear-fusion-project-faces-new-delay-cost-overrun-les-echos-2016-5. Published May 2, 2016.

[92] European Commission. (2010). *ITER & Fusion Research MEMO/10/165*. Press Release. http://europa.eu/rapid/press-release_MEMO-10-165_en.htm?locale=en.

[93] 50 Hertz. (2016). 'Wind Power Data'. *50 Hertz*. http://www.50hertz.com/en/Grid-Data/Wind-power.

[94] Parkinson, G. (2015). 'Grid Operator: 70% solar + Wind on German Grid Before Storage Needed'. *Cleantechnica*. Published December 2015. http://cleantechnica.com/2015/12/11/grid-operator-70-solar-wind-on-german-grid-before-store-needed/.

[95] Neslen, A. (2016). 'Denmark Broke World Record for Wind Power in 2015'. *The Guardian*. https://www.theguardian.com/environment/2016/jan/18/denmark-broke-world-record-for-wind-power-in-2015. Published January 18, 2016.

[96] Appunn, K. (2016). 'Germany's Energy Consumption and Power Mix in Charts'. *Clean Energy Wire*.:1. Published June, 2016. https://www.cleanenergywire.org/factsheets/germanys-energy-consumption-and-power-mix-charts.

[97] Morris, C. (2015). 'Is Germany Reliant on Foreign Nuclear Power?' *Energy Transit Ger Energiewende*. Published June, 2015. http://energytransition.de/2015/06/is-germany-reliant-on-foreign-nuclear-power/.

[98] EPEX Spot. (2016). *European Power Exchange*. https://www.epexspot.com/en/.

[99] GE Power. (2016). *Powering the World with Gas Power Systems*.

[100] Parsons, Brinckerhoff. (2008). 'Cost Estimates for Thermal Peaking Plant "Final Report"'. *Parsons Brinckerhoff New Zeal Ltd Work Pap*. Published June, 2008.

[101] Appunn, K., S. Amelang. (2015). 'Germany's New Power Market Design'. *Clean Energy Wire*. Published June, 2015. https://www.cleanenergywire.org/factsheets/germanys-new-power-market-design.

[102] 50 Hertz. (2016). *Annual Report 2015: The Energy Transition and Its Impact on the 50 Hertz Area*. file:///C:/Users/Rembrandt/Downloads/50Hertz_THE_ENERGY_TRANSITION.pdf.

[103] 50 Hertz. (2016). 'Measures and Adjustments Performed to Meet the Responsibility for the System'. *50 Hertz*. http://

www.50hertz.com/en/Grid-Data/Market-related-measures. Published 2016.

[104] Metering & Smart Energy International. (2013). '1325 MW Energy Storage Target Proposed for California by 2020'. *Metering Smart Energy Int*. Published September, 2013. http://www.metering.com/1-325mw-energy-storage-target-proposed-for-california-by-2020/.

[105] Rehman, S., L.M. Al-Hadhrami, M.M. Alam. (2015). 'Pumped Hydro Energy Storage System: A Technological Review'. *Renew Sustain Energy Rev*. 2015;44(APRIL):586-598. DOI:10.1016/j.rser.2014.12.040.

[106] Guo, A. (2016). 'China's State Grid to Boost Spending Plan 28% to 350 Billion USD'. *Bloomberg*. http://www.bloomberg.com/news/articles/2016-01-21/china-s-state-grid-to-boost-spending-plan-28-to-350-billion. Published January 21, 2016.

[107] GEIDCO. (2016). 'International Conference on Global Energy Interconnection Opens in Bejing'. *GEIDCO*. http://geidca.com/html/qqnycoen/col2015100728/2016-06/01/20160601101607859323413_1.html. Published March 31, 2016.

[108] DNV-GL. (2015). 'Interconnecting the World'. Interview with Liu Zhenya in *Next Sustainable Business*. http://globalcompact15.org/interviews/liu-zhenya.

[109] Producing Food and Fuel. (2015). *Sugarcane.org*. http://sugarcane.org/sustainability/producing-food-and-fuel.

[110] Assunção, J., J. Chiavari. (2006). *Towards Efficient Land Use in Brazil*.:1-28.

[111] Stock, J.H. (2015). *The Renewable Fuel Standard: A Path Fow*.

[112] Wisner, R. (2015). *U.S. Ethanol Usage Projections & Corn Balance Sheet* (*Mil. Bu.*).

[113] Eurostat. (2015). 'Farm Structure Statistics'. *Eurostat Statistics Explained*. http://ec.europa.eu/eurostat/statistics-explained/index.php/Farm_structure_statistics.

[114] European Commission. (2015). 'Bioenergy Technial Background'. *Research & Innoviation Energy*. http://ec.europa.eu/research/energy/eu/index_en.cfm?pg=research-bioenergy-background#future.

[115] Simon, F. (2016). 'Green Transport Target Will be Scrapped Post-2020 EU Confirms'. *Euractiv*. Published May 9, 2016.

[116] Bloomberg New Energy Finance. (2016). 'Biofuels Investment Falls Sharply Since Its 2008 Peak'. *Bloomberg Brief*. Published August 22, 2016.

[117] Dupont. (2015). 'Dupont and New Tianlong Industry Co. Ltd. Sign Historic Deal to Bring Cellulosic Ethanol Technology to China'. *Press Release*. http://www.dupont.com/corporate-functions/media-center/press-releases/dupont-NTL-sign-historic-deal-cellulosic-ethanol-tech-china.html. Published July 16, 2015.

[118] IEA. (2016). *World Energy Investment 2016 Fact Sheet*.

[119] Liebreich, M. (2016). *Bloomberg New Energy Finance Summit*.

[120] World Bank. (2016). 'Gross Domestic Product 2015'. *World Dev Indic database*. 2016;(July):120-123.

[121] IEA: Directorate of Global Energy Economics. (2015). *World Energy Outlook*. DOI:10.1787/weo-2014-en.

[122] Bloomberg New Energy Finance. (2016). *New Energy Outlook 2016 Executive Summary*.

[123] IEA. (2015). *World Energy Outlook 2015*.

[124] González, P. del Río. (2008). 'Ten Years of Renewable Electricity Policies in Spain: An Analysis of Successive Feed-in Tariff Reforms'. *Energy Policy*. 2008;36(8):2907-2919. DOI:10.1016/j.enpol.2008.03.025.

[125] IEA. (2015). *Spain 2015*.: 178.

[126] Paper, E.W., A. Mahalingam, D.M. Reiner. *Energy Subsidies at Times of Economic Crisis: A Comparative Study and Scenario Analysis of Italy and Spain*.

[127] United States Geological Survey. (2013). *USGS National Assessment of Oil and Gas Resources Update Continuous Gas Resources*. DOI:10.1017/CBO9781107415324.004.

[128] USGS. (2016). 'USGS Estimates 66 Trillion Cubic Feet of Natural Gas in Colorado's Mancos Shale Formation'. *Press Release*. https://www.usgs.gov/news/usgs-estimates-66-trillion-cubic-feet-natural-gas-colorado-s-mancos-shale-formation. Published June 8, 2016.

[129] Curtis, J.B., R.J. Kelley, S.M. Hamburg, N.H. Reagan. (2015). *Potential Supply of Natural gAs in the United States*. DOI:10.1017/CBO9781107415324.004.

[130] Gough, P.J. (2016). 'Lacking Infrastructure Creates Vast Inventory of Uncompleted Wells'. *Pittsburgh Business Times*. http://www.bizjournals.com/pittsburgh/print-edition/2016/06/10/lacking-infrastructure-creates-vast-inventory.html. Published June 10, 2016.

[131] Williamson, R. (2016). 'Shell and Apache Push for Egyptian Tight Gas'. *Interfax*. http://interfaxenergy.com/gasdaily/article/20695/shell-and-apache-push-for-egyptian-tight-gas. Published June 16, 2016.

[132] Aboudi, S., D. Fineren. (2013). 'BP Signs 16 Billion USD Tight Gas Project Deal in Oman'. *Reuters*. http://www.reuters.com/article/us-bp-oman-gas-idUSBRE9BF0DQ20131216. Published December 16, 2013.

[133] British Petroleum. (2016). 'BP Deepends Commitment to Oman; to Extend Licence and Develop Second Phase of Major Khazzan Gas Gield'. *Press Release*. http://www.bp.com/en/global/corporate/press/press-releases/bp-deepens-commitment-to-oman.html. Published February 14, 2016.

[134] Virginia Department of Mines Minerals and Energy. (2015). *Coal Deposits in Virginia.* https://www.dmme.virginia.gov/dgmr/coal.shtml.

[135] West Virginia University. (2015). 'WVU Report Shows Coal Industry Faces Nearly 39 Percent Decline in Coal Production'. *Press Release*. http://wvutoday.wvu.edu/n/2015/05/28/wvu-report-shows-coal-industry-faces-nearly-39-percent-decline-in-coal-production. Published May 28, 2015.

[136] U.S. Energy Information Administration. (2016). *EIA Weekly Coal Production by State*. http://www.eia.gov/coal/production/weekly/. Published August 4, 2016.

[137] U.S. Energy Information Administration. (2010). *Annual Coal Report 2008*. Vol 0584.

[138] EIA. (2013). *Annual Coal Report 2013*.

[139] U.S. Energy Information Administration. (2010). *Annual Coal Report 2010*. Vol 0584.

[140] U.S. Energy Information Administration. (2016). *Annual Coal Report 2013*.

[141] Virginia Centre for Coal and Energy Research. (2016). *Virginia Energy Patterns and Trends*. Webpage. https://www.energy.vt.edu/vept/coal/.

[142] Williams, B.A., M.J. Cassidy. (2016). 'Coal Tax Credits Aren't Working'. 2016;(March):1-2.

[143] Zipper, C.E. (1996). *Effects of Virginia Coalfield Employment Enhancement Tax Credit Legislation*.

[144] IEA, (2016). *Energy and Air Pollution*.

[145] Enerdata. (2016). 'World Power Plant Database'. http://www.enerdata.net/enerdatauk/knowledge/subscriptions/research/power-plant.php.

[146] End Coal. (2016). 'EndCoal Power Plant Tracker'. *Webpage*. http://endcoal.org/global-coal-plant-tracker/.

[147] U.S. Environmental Protection Agency. (2015). *Carbon Pollution Emission Guidelines for Existing Stationary Sources: Electric Utility Generating Units*. Vol 80.

[148] Anonymous Nuclear Experts. (2015). 'Reactivity Coefficients: Reactivity Feedbacks'. *nuclear-power.net*. http://www.nuclear-power.net/nuclear-power/reactor-physics/nuclear-fission-chain-reaction/reactivity-coefficients-reactivity-feedbacks/.

[149] Bradsher, K. (2011). 'A Radical Kind of Reactor'. *The New York Times*. http://www.nytimes.com/2011/03/25/business/energy-environment/25chinanuke.html. Published March 24, 2011.

[150] Halper, M. (2012). 'What to do With 135,000 Pebbles: Generate a Lot of CO2 Free Safe Nuclear Power, Says South African Startup'. *Weinberg Next Nuclear*. http://www.the-weinberg-foundation.org/2012/10/12/what-to-do-with-135000-pebbles-generate-a-lot-of-co2-free-safe-nuclear-power-says-south-african-startup/. Published October 12, 2012.

[151] Zhang, Z., Z. Wu, Y. Xu, Y. Sun, F. Li. (2004). 'Design of Chinese Modular High-temperature Gas-Cooled Reactor HTR-PM'. *2nd International Topical Meeting on High Temperature Reactor Technology*.:1-8. http://citeseerx.ist.psu.edu/viewdoc/download?doi=10.1.1.176.1632&rep=rep1&type=pdf.

[152] Abdulmohsin, R., A. Dissertation. (2013). *Gas Dynamics and Heat Transfer in a Packed Pebble- Bed Reactor for the 4 Th Generation Nuclear Energy.*

[153] The Economist Staff. (2002). 'Pebble Dashed?'. *The Economist.* http://www.economist.com/node/1200601. Published June 27, 2002.

[154] Next Kraftwerke. (2015). 'Secondary Reserve & Secondary Control Power'. *Next Kraftwerke.* https://www.next-kraftwerke.de/wissen/regelenergie/sekundaerreserve.

[155] Engelmair, R. (2016). 'Ausschreibungen der Primarregelreserve in der Regelzone APG'. *APG.* https://www.apg.at/de/markt/netzregelung/primaerregelung/ausschreibungen.

[156] Consentec GmbH. (2014). 'Description of Load-Frequency Control Concept and Market for Control Reserves'. 2014;(February):43. http://www.consentec.de/wp-content/uploads/2014/08/Consentec{_}50Hertz{_}Regelleistungsmarkt{_}en{_}{_}20140227.pdf.

[157] Fialka, J. (2016). 'World's Largest Storage Battery Will Power Los Angeles'. *Sci Am.* July 2016. http://www.scientificamerican.com/article/world-s-largest-storage-battery-will-power-los-angeles/.

[158] Hruska, J. (2014). 'World's Largest Lithium-Ion Battery to be Built in Southern California, Dwarfs Previous Installations'. *ExtremeTech.* Published November, 2014.

[159] O'Connell, K., D. Pacini. (2015). 'LG Chem to Supply Battery Modules for AES Energy Storage Advancion Solution'. *LG Chem Press Release.* http://lgcpi.com/2015/12/15/lg-chem-to-supply-battery-modules-for-aes-energy-storage-advancion-solution/. Published December 15, 2015.

[160] John, J. (2016). 'Eos Energy Storage Is Raising $23M to Scale Up Zinc-Based Grid Battery Production'. *Green Tech Media.* https://www.greentechmedia.com/articles/read/eos-energy-storage-is-raising-23m-to-scale-up-zinc-based-grid-battery-produ. Published October 19, 2016.

[161] Phillips, L. (2016). 'Siemens Tests Breakthrough Basalt Rock Thermal Storage System for Excess Power Generated by Wind Power Systems'. *District Energy.* http://www.districtenergy.org/blog/2016/10/12/siemens-tests-breakthrough-basalt-

rock-thermal-storage-system-for-excess-power-generated-by-wind-power-systems/. Published October 12, 2016.

[162] China National Energy Board. (2014). National Energy Board Released the Total Electricity Consumption in 2013. Published January 14, 2014.

[163] China National Energy Board. (2015). National Energy Board released the total electricity consumption in 2014. Published January 16, 2015.

[164] CCTV.com. (2016). 'North China Wind Turbines Face Oversupply Issues'. *CCTV.com.* http://english.cctv.com/2016/04/08/VIDEJ6Q5RSMNXKn1CO4sCQL7160408.shtml.

[165] Power Engineering International. (2016). 'China Orders Grid Operators to Connect Renewables. *Power Eng Int.* Published March, 2016. http://www.powerengineeringint.com/articles/2016/03/china-orders-grid-operators-to-connect-renewables.html.

[166] Siemens. (2014). *Fact Sheet High-Voltage Direct Current Transmission (HVDC).*

[167] ABB. (2014). *Special Report 60 Years of HVDC.*

[168] Converter Factory. (2016). 'The Energy Transition in Germany: Siemens Supplies Converters for Grid Expansion to Amprion and Transnet BW PK'. *Converter Factory.* Published April 8, 2016.

[169] Radowitz, B. (2015). 'Germany Goes Underground with HVDC Transmission Lines'. *RECHARGE.* http://www.re-chargenews.com/wind/1413351/germany-goes-underground-with-hvdc-transmission-lines. Published October 7, 2015.

[170] Tamiro, R. (2011). 'Going Underground: European Transmission Pracices'. *Electric Light & Power.* http://www.elp.com/articles/powergrid_international/print/volume-16/issue-10/features/going-underground-european-transmission-practices.html. Published January 10, 2011.

[171] ABB. (2014). 'ABB Develops Complete System Solution for 1100 kV HVDC Power Transmission'. *ABB.* http://www.abb.com/cawp/seitp202/d83ce03a99b85dd7c1257d400041aab5.aspx. Published August 26, 2014.

[172] Eisentraut, A., A. Brown, L. Fulton, et. al. (2011). *Biofuels for Transport in 2050.* DOI:10.1002/bbb.330.

Register